STELLINGEN.

I

De hypothese van Spurr omtrent de ontstaanswijze van lensvormige, hypogene ertslichamen, die geen zichtbaar voedingskanaal bezitten, is verwerpelijk; hun ontstaan kan alleen door metasomatose verklaard worden.

II

De hypothese van L. C. Graton omtrent een hypogene oorsprong der Witwatersrand-goud-afzettingen is aannemelijk.

III

Bij tinertsen van het Ankole type kan door getrapt breken een concentratie bewerkt worden, die een eventueel vervoer naar een Centrale wasscherij belangrijk goedkooper maakt en waardoor tafelscheiding overbodig wordt.

Uitzondering moet worden gemaakt met erts uit sterk gesilificeerde ganggedeelten.

IV

De asymmetrische vorm, welke een profiel over de Edward-Rift te zien geeft, kan het best verklaard worden, door een plooibrekende opschuiving aan te nemen, met een horizontale verplaatsing van West naar Oost.

V

Alleen wanneer in een sub-aried gebied een steilwand geen symmetrisch gelegen tegenhanger heeft, is het vermoeden gerechtvaardigd dat deze van tectonischen oorsprong is.

VI

Indien in een sub-aried gebied twee symmetrisch tegenover elkaar liggende steilwanden voorkomen, is het vermoeden, dat zij van tectonischen oorsprong zijn, alleen dan gerechtvaardigd, wanneer bewezen kan worden, dat het tusschenliggende dal niet door rivier-erosie kan zijn gevormd.

VII

De meening van E. O. Meyer, dat zijn kaarteering van de „Grosze Bruch-stufe" geen correctie meer behoefde, is ongetwijfeld onjuist (Verg. N. Jahrbuch für Min. Beilage Bd. 38, S. 807).

VIII

Men dient aan te nemen, dat de Karagwe-Ankole granieten niet zijn ontstaan door samensmelting van kwartsrijke sedimenten en een primair basisch magma, doch dat het magma reeds bij aanvang der intrusie een granitische samen-stelling had.

IX

Taakwerk is voorloopig de eenig mogelijke arbeidsvorm voor Centraal Afrikaansche negers.

X

Bij een eventueele exploitatie van tinmijnen in Z.W. Uganda zal men er rekening mee moeten houden, dat het kweeken van een kern van geschoolde arbeiders op zeer groote bezwaren zal stuiten.

XI

Het is van groot belang bij een eventueele exploitatie van tinmijnen in Z-W. Uganda de belangrijkste lucht- en transportgangen zoo snel mogelijk te torkreteeren, te meer omdat het betrekken van mijnhout zeer groote moeilijk-heden oplevert.

XII

Minder modern uitgeruste smelters dienen er hun aandacht aan te wijden, dat alluviale concentraten zoodanig verontreinigd kunnen zijn door markasiet, dat een voorafgaande roosting noodzakelijk wordt.

XIII

De naam "Tor", door Gregory voorgesteld voor zuiver consequente rest-heuvels in een sub-aried gebied, is verwerpelijk; de naam "Inselberg", of "island mount" dient gehandhaafd te blijven.

XIV

Voor vele vergrijpen, gepleegd door inboorlingen van Centraal Afrika zijn lijfstraffen de meest billijke straf.

XV

Aan een werkgever in Centraal Afrika dient een, zij het begrensde, strafrechterlijke bevoegdheid gegeven te worden, in het bijzonder wanneer hij zich op grooten afstand van een zetel van het Europeesch Bestuur bevindt.

XVI

De werkgever dient in Centraal Afrika de bevoegdheid te bezitten, om aan een werknemer, wiens prestaties onvoldoende waren, het bewijs te onthouden dat door dien werknemer een dagtaak is verricht.

THE GEOLOGY OF SOUTHWESTERN UGANDA

THE GEOLOGY

OF

SOUTHWESTERN UGANDA

WITH SPECIAL REFERENCE TO THE STANNIFEROUS DEPOSITS

PROEFSCHRIFT

TER VERKRIJGING VAN DEN GRAAD VAN DOC-
TOR IN DE TECHNISCHE WETENSCHAP AAN DE
TECHNISCHE HOOGESCHOOL TE DELFT, OP GE-
ZAG VAN DEN RECTOR MAGNIFICUS, Ir. H. TER
MEULEN, HOOGLEERAAR IN DE AFDEELING
DER SCHEIKUNDIGE TECHNOLOGIE, VOOR
EEN COMMISSIE UIT DEN SENAAT TE VER-
DEDIGEN OP WOENSDAG 17 FEBRUARI 1932,
DES NAMIDDAGS TE 4 UUR

DOOR

HENDRIK ALBERT STHEEMAN

MIJNINGENIEUR

GEBOREN TE TETERINGEN

SPRINGER-SCIENCE+BUSINESS MEDIA, B.V.

ISBN 978-94-017-0021-4   ISBN 978-94-015-7553-9 (eBook)
DOI 10.1007/978-94-015-7553-9

VOORWOORD

Met dit proefschrift wordt een episode van mijn leven afgesloten.

In de eerste plaats valt daarin mijn studietijd aan de Technische Hooge-school en groot is mijn erkentelijkheid voor het vele goede, dat ik daar heb mogen ontvangen. In het bijzonder ben ik aan U zeer veel verplicht, Hoog-geachte Promotor Grondijs. U hebt mijn eerste schreden geleid in de studie der mineragraphie en den leer der ertsafzettingen. Daarnaast zal ik Uw vriend-schappelijke omgang met Uw studenten blijven gedenken. Door U heb ik deel kunnen nemen aan de exploratie in Uganda en na mijn terugkomst hebt U mij, als mijn toekomstige Promotor op alle mogelijke wijzen gesteund.

U, Hooggeleerde Molengraaff mocht ik niet meer terugvinden als hoog-leeraar. Uwe belangstelling in het werk Uwer leerlingen was echter geenszins gedoofd! Ik hoop, dat U in dit proefschrift, dat handelt over een continent, waaraan zoo'n belangrijk deel van Uw levenswerk was gewijd, iets van de objec-tiviteit moogt terugvinden, waarvan Uw gloedvolle colleges getuigden. Zoo ik heb ervaren, dat de liefde en de bewondering voor de natuur stijgt bij een begrijpend aanschouwen, zoo is het zaad daarvoor grootendeels door U gelegd.

Ook U, Hooggeleerde Brouwer mocht ik bij mijn terugkomst in Delft niet meer aantreffen. Uw koel en nuchter oordeel heb ik mij vaak in herinnering gebracht bij het overpeinzen van geologische problemen; Uw enthousiasme en onvermoeid doorzettingsvermogen waren mij een aansporing op mijn geolo-gische tochten.

U, Hooggeleerde Grutterink dank ik voor Uw onderwijs en voor Uw bereid-willigheid, wanneer ik Uw hulp of advies inriep.

Uw onderwijs, Hooggeleerde van Nes en Caron heeft mij ten slotte tot inge-nieur gevormd. U beiden hebt op mij overgedragen den drang om moeilijk-heden, die de natuur ons in den weg legt, op economische wijze te over-winnen.

Bij U, Hooggeachte Promotor Mekel, heb ik geen onderwijs genoten in den engeren zin des woords, doch hoeveel heb ik van U geleerd in dezen tijd, dat ik met U in aanraking kwam! Van Uw steun en medewerking kon ik steeds verzekerd zijn en ondanks Uw druk bezette tijd waart U steeds tot hulp bereid.

Door Uw klaar en helder woord en Uw opbouwende kritiek hebt U, evenals Professor Grondijs, bijzonder veel bijgedragen tot verheldering van dit ge-schrift en ik dank mijn beide Promotoren voor de even openhartige, als wel-willende wijze, waarop zij mij steeds hun oordeel te kennen gaven.

Met dit proefschrift wordt echter niet slechts mijn studietijd afgesloten. Het is tevens een afscheid aan een betrekking, aan een vak, een land.

Het sluit een periode af, waarin ik onder Uw leiding, Hooggeachte Heer de Greve, in dienst van de Centraal Afrikaansche Exploratie Maatschappij zoo buitengewoon prettig heb mogen werken. Van U kwam de eerste aansporing om mijn werk in Uganda voor een dissertatie te benutten en van U kwam de eerste steun, die daarvoor noodzakelijk was.

U, Hooggeachte Heer van den Broek, hebt mij alle steun en medewerking doen geworden, die voor het volbrengen van dit werk noodzakelijk waren, zoowel finantieel, als op andere wijze. Met buitengewone liberaliteit hebt U mij toegestaan mijn gegevens te publiceeren en mij in staat gesteld deze met gegevens van de Maatschappij tot een overzichtelijk geheel af te ronden. Uw bereidwilligheid, wanneer ik Uw steun inriep, zal mij steeds in herinnering blijven, doch bovenal zal ik de persoonlijke moeiten gedenken, die U zich als mijn vroegere Directeur hebt getroost, om mij in deze tijden weer een betrekking te verschaffen.

U, Hooggeachte Heer van der Linden, dank ik voor Uw persoonlijke belangstelling en Uw waarlijk vriendschappelijke raadgevingen.

Aan de ertsgeologie, zoowel als aan Afrika een laatst vaarwel. Het eerste had mijn volle toewijding, en van Afrika, het land van vergezichten en kleuren, en bovenal van het leven in Afrika, ben ik gaan houden, zooals een ieder, die daar gewoond heeft.

Dat ik mijn herinneringen heb mogen uitstorten in dit boek en op deze wijze van land en vak heb afscheid mogen nemen, maakt dit afscheid zeer veel lichter.

Hen, die mij geholpen hebben tijdens het schrijven van dit proefschrift, heb ik elders mijn dank gebracht. Hier moet ik nog gewagen van de belangstelling van den Heer van Doorninck, met wien ik enkele malen op vruchtbare wijze van gedachten mocht wisselen, en de prettige samenwerking met den Heer Speyer in Aug.-Sept. 1930 te Kaina.

Het enthousiasme van mijn medewerkers, de Heeren Oosterchrist en Provily, heb ik steeds op bijzonder hoogen prijs gesteld.

This book is dedicated
to my Wife,
my Comrade in Africa.

CONTENTS

Page

PREFACE XI
ACKNOWLEDGEMENTS XVII

PART I — MORPHOLOGY

1. GENERAL INTRODUCTION 1
2. THE VALLEY FLOOR, CREEP AND FLOOD SHEET 4
3. VALLEY SIDES — FORMATION OF ARENAS 10
4. CONCENTRATING POWER OF THE RIVERS 17
5. EVIDENCES OF WARPING — SWAMP DIVIDE 20

PART II — REGIONAL GEOLOGY

1. WISHIKATWA 25
2. DEVELOPMENT OF THE LOWER DIVISION OUTSIDE WISHIKATWA . . . 29
3. MIDDLE AND UPPER DIVISIONS OF THE KARAGWE-ANKOLIAN SYSTEM . 33
4. A TECHNICAL DISCUSSION OF THE MAPPING OF ARENAS 40
5. CORRELATION 43
6. DESCRIPTION OF THE GRANITES — ABNORMAL TEXTURES — BASIC DYKES 46

PART III — TECTONICS

1. VARIOUS PHASES OF FOLDING — NATURE OF CONTACT — MANNER OF
   INTRUSION 53

PART IV — METAMORPHISM
A. AUTOMETAMORPHISM

1. INTRODUCTON 65
2. INJECTION SCHISTS 69
3. THE FORMATION OF MAGMATIC SOLUTIONS 72

B. ALLOMETAMORPHISM CAUSED BY MAGMATIC SOLUTIONS

4. INTRODUCTION; SEQUENCE OF DEPOSITION 79
5. SERICITE PHYLLITES IN THE VICINITY OF GRANITE 82
6. METAMORPHISM ALONG FAULTS — FORMATION OF METASOMATIC VEINS 87

PART V — THE STANNIFEROUS DEPOSITS

1. INTRODUCTION 98

A — KAINA

2. GENERAL DESCRIPTION 106
3. MICROSCOPICAL INVESTIGATIONS 114

B. STANNIFEROUS DEPOSITS IN PREDOMINANTLY ARENA-CEOUS ROCKS

4. GENERAL INTRODUCTION 122
5. TRANSITIONAL ZONE OF KASHOZO VEINS 127
6. THE VEIN-FILLING — BURAMMA AND RUECHIMARRA DEPOSITS . . 130
7. CRUSHED QUARTZ 134

C. STANNIFEROUS DEPOSITS IN THE VICINITY OF GRANITE

8. GENERAL DESCRIPTION 137
9. MICROSCOPICAL INVESTIGATIONS 139

D. CONCLUSIONS
142

MAPS AND SECTIONS:

Map 1. Map of Uganda Protectorate, after map of Uganda Survey Department No. A 530 24
Sections through Lower Division of Karagwe-Ankolian System 32
Map 2. Situation of Karagwe-Ankolian Outcrop in Central Africa 52
Map 3. Survey Map of SW. Uganda and Adjoining Territories; scale 1 : 500,000 64
Map 4. Geological Map of Rushenyi, Wishikatwa, adjoining Kigezi and Kavungo; scale 1 : 100,000 . . . . . . Appendix 1
Map 5. Sketch Map of Katuba Valley and Wishikatwa; scale approx. 1 : 30,000 . . . . Appendix 2
Map 6. Sketch Map of part of Kigezi; scale 1 : 100,000 . . Appendix 3

N.B. The numbers of slides and hand-specimens are those of the collection handed to the Mining Institute of the Technical University at Delft. The illustrations of parts of slides are of different magnifications; generally the actual diameter of the parts illustrated is 2—3 mm; the diameter of figs. 42, 47, 64, 85, 88, 89, 90 is much smaller, about $1/2$—$1^1/2$ mm.

PREFACE

In order that the reader may properly judge the scope of the information given in this paper it is necessary to explain the manner in which it has been compiled and the facts which have led to its publication. It is the fruit of eighteen months' geological exploration and prospecting in the service of the Central African Exploration Company (a subsidiary of the Billiton Company), which in co-operation with a British company acquired large concessions in southwestern Uganda and in northwestern Karagwe.

Uganda, which together with Kenya Colony and Tanganyika Territory[1]) forms part of British East Africa, has an area of 244,000 square kilometres, of which 42,700 sq.km. is water, including the northern part of Victoria Nyanza with an area of 67,000 sq.km. This protectorate (cf. map 1, drawn after the map 1 : 1,000,000 of Uganda Survey Dept. No. A 530) comprises three kingdoms, Buganda, Toro and Ankole, and some Crown lands divided into districts, the southwesternmost of which, Kigezi, with Kabale as its seat of European Administration, will be frequently mentioned.

The first European to enter Uganda was Captain Speke, who reached Gondokoro in 1862 just as Sir Samuel Baker was preparing for his journey to the south. The names of these two explorers recall the times of arduous pioneer work in discovering the sources of the Nile, when such magnificent examples of courage and perseverance were shown and attention was drawn to those remote parts of Africa known to-day as British East Africa.

Soon afterwards British and French missionaries started to traverse Central East Africa, gathering detailed information about the people and the country.

Little was known at that time of the geology of those parts, though the descriptions of volcanoes and big lakes given by the early travellers aroused some interest. More detailed and reliable geological information did not, however, begin to become known until the rising German Empire turned its attention to Tanganyika, which was visited by scientific missions that did splendid work and contributed in no small degree to the store of geological knowledge; in those days the minds of explorers were mostly occupied with the younger geological history of Africa, especially with the evidences of comparatively recent volcanism and the formation of enormous structural valleys.

---

1) This is the name given to the greater part of what was formerly German East Africa, which is now administered by Great Britain under a mandate from the League of Nations; other parts of the former German territory, Ruanda and Urundi, are under the mandatory administration of Belgium.

In his "Antlitz der Erde", Suesz gave a description covering the whole of the African fault system, since when this problem has been the subject of many papers. Suesz believed the Rift Valleys to have been formed in comparatively recent time, but Gregory demonstrated the complicated nature of the formation of those enormous subsidences. Though there are still many gaps to be filled and doubtless his explanation does not hold good for all African structural valleys, it is much to Gregory's credit that the existence of alternating phases of volcanic activity and tectonic movement has been proved in the eastern branch of the Rift Valley. Kenya alone, however, cannot furnish the solution of the whole of the Rift Valley problem, there being considerable differences between the eastern and the western branch, on the borders of which latter branch Uganda lies.

This western branch of the Rift Valley system has been particularly explored by Mr E. J. Wayland, who followed up very closely the various movements connected with the formation of the Albertine Rift.

The various old formations which build up Central Africa proving to be rich in useful minerals, much attention was paid to its older geological history. The Belgians found their colony to be exceedingly well favoured by nature and they displayed great activity in its development. Great Britain did not lag behind, and indeed every geologist working in Uganda to-day is greatly indebted to the Uganda Geological Survey Department for the work it has done. Before any ore was known to occur, certain localities were marked as being more or less likely to yield minerals, and in fact the Survey Department has greatly contributed, within the limits of its sphere of operation, to the economic development of the country by controlling the water supply and assisting prospectors by narrowing down the likely areas, for which purpose several parts have been mapped in detail.

The outcrop of the Karagwe-Ankolian System in the SW. of Uganda was considered to be very promising. This system was noticed by Captain Speke already in 1861 both in Karagwe and in Ankole, and it was first described as the Karagwe System by J. W. Gregory and Scott Elliot in 1895.

The Geological Survey Department rightly interpreted the presence of granites and pneumatolytic minerals as a favourable indication, and in the years 1922 to 1926 the Field Geologist A. D. Combe undertook the mapping of certain parts of Ankole, in which work he was greatly assisted by the carefully compiled topographical map made by the Anglo-German Boundary Commission (1905). Mr Combe's geological map appeared in Central Africa in January 1930 and has proved to be of immense value for prospecting work.

Cassiterite was first discovered in 1924 in Karagwe, and many companies were formed for its exploitation, but apart from the Mwirisandu vein and the alluvial deposit near Kyerwa no paying deposits were discovered. The possibility of rich deposits being found was not, however, precluded, and by following a wise policy in regard to concessions the Government succeeded in attracting some financially strong companies. Thus it came about that in 1929 the Billiton Company sent out an expedition in search of tin.

The present paper is the result of the author's work for the C.A.E.C. and it must be emphasized that, though this is primarily a dissertation for taking the degree of Doctor in the technical sciences, the information on which it is based was not compiled for a purely scientific purpose.

The author had to meet the cold and clear-cut demands of practice by applying the scientific grounding that he had received at the Technical University of Delft, and his foremost duty was to trace the conditions essential for the occurrence of primary or secondary stanniferous deposits. Consequently some regions have necessarily been more thoroughly explored than others, while also certain subjects have received less attention than would otherwise have been the case in a purely scientific exploration. Physiography, for instance, has been dealt with in much more detail than is usually the case, for the reason that in the effect of the sub-arid cycle lies the key to the prospector's work. A close study of the physiography was necessary not only to determine where the chances of finding alluvial tin deposits exist, but also to estimate the value and location of the primary source from discoveries of secondary tin. Stratigraphy, on the other hand, has on the whole received rather scanty consideration, except for that of the lower Karagwe-Ankolian division, which has been examined very carefully, for it is in this division that by far the most of the known primary tin deposits occur, whereas the areas built up exclusively from the upper horizons lack every indication of the presence of tin.

This being confirmed by an investigation of the structure, special attention was paid to the tectonics, at least in so far as the results might be of importance for the discovery of tin veins. Consequently the structure of Kigezi has been studied only on general lines, whilst that of the areas in the vicinity of the granite bodies has been very minutely explored.

The relations between the Karagwe-Ankolian system and other systems, either younger or older, have, of course, not been traced.

Undoubtedly the description of the normal granites given hereinafter is very incomplete, but full attention has, however, been paid to alterations in the granite, or, one might say, to symptoms relating to the genesis of tin veins. In consequence a great deal of the work herein described has been devoted to phenomena of autometamorphism and the description of pegmatitic varieties of the granite.

Contact and dynamo-metamorphism have been described on only broad lines, except where important results were to be reported that were likely to throw some light on the structure of the country or the genesis of cassiterite. Obviously, however, the metamorphism commonly called pneumatolytic has been given a very special place in this paper, as also the composition and shape of the tin veins and the sequence of deposition of the constituent minerals. At the same time an attempt has been made to trace the mechanism of the ore deposition right from the beginning. Many criteria for estimating the value of veins based on experience alone have been confirmed by the results of this investigation and given logical grounds.

Thanks to the information contained in publications by Combe and Salée and the data collected by members of the staff of the C.A.E.C., it has been possible for the author to show something of the relation existing with the surrounding regions and also to compile a survey map (No III) and to complete two other maps (IV and VI).

The present paper deals in detail with the following:

1. The stratigraphy and tectonics of the Lower Division of the Karagwe-Ankolian system in so far as this occurs in the districts of Rushenyi and Wishikatwa, with the adjacent Kigezi (this part is also shown on Combe's map mentioned above).

2. The stratigraphy and tectonics of Kigezi, on which no data had hitherto been collected. In this district it is particularly the Middle and Upper Divisions that are exposed.

3. The autometamorphic alterations in the granite and the sequence of deposition of the various metasomatic minerals.

4. The allometamorphism due to peripheral magmatic emanations containing barium and fluorite. It will be seen that the sequence of deposition of the metasomatic minerals in the sediments is very similar to that of the contact-metamorphic minerals in the granite.

5. The shape, occurrence and composition of the stanniferous deposits and their metamorphic aureoles, the sequence of deposition of which again appears to be very similar to that of the above-mentioned metasomatic minerals. Moreover it will be shown that the genesis of the stanniferous deposits in Ankole is due to very intensive metasomatism. On the whole a fairly detailed picture could be drawn of the pathways followed by magmatic emanations through the Karagwe-Ankolian sediments, and their effect therein.

The writer was especially charged with the prospecting for primary deposits, but the detrital shows of cassiterite connected with these veins yielded such great difficulties that his attention was naturally drawn to the problem of sub-arid erosion, and it is for this reason that such an important place is given to morphology.

His explorations not having been particularly directed towards the tracing of alluvial deposits, the writer has not gone very deeply in this paper into the development of Uganda's river system. Only twice has this subject been touched upon, and for two reasons. In the first place because it had been noticed that only a few rivers lend themselves to the formation of alluvial deposits, apart from the geological nature of the areas through which they flow, for in the majority of the valleys a running river has long ceased to exist, and now only the ground water slowly percolates downwards, the thick layer of detritus creeping towards the actual river, where it is carried along farther by running water; only in these ultimate rivers is a concentration of specifically heavy and useful minerals possible, thanks to the sorting effect of the running water. Another reason — and this explains why the river system of Kigezi has been specially dealt with — is that in the course of the rivers and the trend of the old peneplain (still clearly to be traced on the mountain ridges) the writer found some evidence for the hypothesis evolved by Wayland, who is of the opinion that the western branch of the Rift Valley has not been formed by the subsidence of a block between two normal fault planes, but rather that its formation was accompanied by a folding of the surface, which might indicate compressional force. The writer is of the opinion that the last word has not been spoken in this connection, since cross-sections made through the Edward Rift clearly show the asymmetric form of this subsidence, so that what is true of the eastern side need not necessarily apply to the western side. Nevertheless, both the topography and the river system of Kigezi give indications that the aforementioned warping does indeed exist on the eastern side.

Further this paper deals extensively with the formation and development of the arenas, large plains consisting principally of granite and encircled by a belt of very steep hills built up from sediments. It is on these arenas, which give such a peculiar character to the region formed by the Karagwe-Ankolian system, that the writer's work has mainly been concentrated. In the beginning attention was paid to the hill slopes, as their thin covering of detritus and the numerous outcrops promised the best possibilities for collecting information. Naturally the author's attention was also turned to the formation and manner of regression of these steep slopes, and it particularly interested him to know whether the origin of the steep wall embracing the arenas was consequential or subsequential. In this he endeavoured to follow Davis' footsteps. It is a good and very fruitful principle to deduce and analyse first the consequences of a certain type of climate, and no less to study the typical differences in morphological features produced by various cycles and correlate these with the characteristic differences shown by the climates themselves. Only when the consequent features have been carefully studied can an attempt be made to apply these consequences to the subsequent features shown by nature. From this investigation it appeared that although in many parts the arena wall coincides with the outcrop of a quartzite and with the contact between granite and sediment, it is not by any means to be concluded that the arena wall is a subsequent formation. The relation between quartzite outcrop and arena wall is purely one of a transient nature, and the steep angle of dip, which is constant from foot to top, is purely a consequence of the sub-arid climate. In this respect the writer's conclusions agree entirely with what W.M. Davis has written in several articles in the Journal of Geology (Jan. and Febr. 1930).

In the latter part of his stay in Uganda the writer had particulary to deal with stanniferous veins and detrital shows of cassiterite on the old peneplain and in the extensive, gently sloping plains called arenas. He believes to have succeeded in solving the problem as to how the detritus from the hill sides is transported across the arena floor to the ultimate draining river. At the same time it has been made quite clear what roles are played by flood sheet and creep in the process of formation of these plains, which in all sub-arid regions cover such enormous areas. In this way it has been possible to form an idea of the formation and development of what is sometimes termed the rock floor but which is here referred to as the arena or valley floor.

Gratitude is due to the management of the Central African Exploration Company not only for its courtesy in allowing the publication of this work but also for the substantial support it has given. Besides the author's own experiences the book contains the general information gathered by the Company's geologists and engineers in their mapping work, and indeed the Company's Management has adopted an extremely liberal attitude in regard to publication, undoubtedly fully recognizing the great importance of the data collected by their geologists at a cost which no academical mission or individual could afford. It would have been all the more regrettable if this information had not been published, since present world conditions will not tend to encourage anyone to undertake in the near future exploration or research work in Uganda on such a large and liberal scale as that organized by the C.A.E.C., so that the continuation of the geological research in Uganda will be left only to the few members of the Geological Department.

Consequently the Management's resolution to allow publication of this work is to be all the more appreciated in view of the fact that the company's mining works have furnished evidence which is still badly needed in even more accessible and civilized parts of the world.

It is for these reasons too that this book has been published in the English language, because were it written in Dutch, the author's own language, the benefit of the results of the work done would be available to only a comparatively few.

A further important consideration is the fact that it concerns a British Protectorate.

In the preface given in the Dutch language the author has already expressed his gratitude to his professors and all who have assisted, in one way or another, in the publication of this work, and he now wishes to say how highly he appreciates the excellent work done by the Geological Department of Uganda under the supervision of the Director, Mr. E. J. Wayland, the results of which have been extremely helpful; he is much appreciative of the opportunity given him to exchange views with the field geologist Mr A. D. Combe, whose map, as soon as it became available in Uganda, made it no longer necessary to continue the mapping of large areas.

Further, a word of sincere thanks is due to the Government Officials for their personal assistance, and in particular to Captain J. E. Philipps, at the time District Commissioner of Kigezi, for his valuable advice and help.

ACKNOWLEDGEMENTS

The author is very much indebted to:

Mr. F. G. GARRATT, who undertook the translation of the Dutch text and the revision of the English text, for his especial zeal and helpfulness. Many an obscure passage and dark point has been made clear as a result of his comments, which speaks for the great attention he has devoted to the work.

Dr. P. KRUIZINGA, for his valuable advice and personal attention to the preparation of the micro-photographs.

Ir. J. DE VRIES m.i., for his readiness to help in determining the minerals whenever the author was in doubt.

Mr. C. VAN WERKHOVEN, who has shown great diligence in preparing maps 1, 2, 3, 4 and 6, which have been made both striking and artistic thanks to his having made himself acquainted as far as possible with the geology of the country. Furthermore he has devoted much care to the making of prints of the photos.

Mr. L. W. N. VAN LEEUWEN, for his scrupulous care in drawing the sections and sketches for Parts I and II and some others.

Finally, the author wishes to thank the publishers, MARTINUS NIJHOFF, for the excellent despatch and particular care bestowed upon the publication of his work.

PART I

MORPHOLOGY

CHAPTER 1

GENERAL INTRODUCTION

A sub-arid climate is characterised by fair'y little precipitation and pronounced dry seasons. It thus gives rise to a typical vegetation, which, as will be seen in the following pages, is the main cause of the exceptional character of the sub-arid form of landscape. The same vegetation is found over large areas in Central Africa in spite of great differences in the amount of rainfall [1]. The reason for this lies in the fact that in one certain area the annual rainfall may vary by as much as one metre or more; in the usual wet season it sometimes occurs that there is no rain at all, while a dry season may be marked by heavy and abundant rains. The vegetation, however, must be such as will only thrive with the minimum of rainfall. These factors, together with the scanty relative precipitation, result in the vegetation consisting mainly of grasses, with in places also some aloe-like plants and thorn-bush. In SW. Ankole in fact grass is the only vegetation. It is not clear whether this is a natural phenomenon or whether it is a consequence of the cattle-driving Bahuma regularly burning down the grass in the dry season, thus preventing the growth of trees [2]. Here the natives burn manure, as is done in Thibet!

The few trees that are found are mainly spiny acacia and candelabra euphorbia (the latter principally on granite soil), both types which can stand long periods of drought. As a matter of fact two sub-types of vegetation are to be distinguished; one, around Lake Victoria and more variegated in tree species, characterises the landscape as a savannah with small scattered woods (bouquets d'arbres) and forest in the river valleys (Martonne's "Soudan" type); the other sub-type occurs in Ankole, a steppe-like tract with thorn-bush and trees regularly and widely distributed (Senegal type — see photo 1).

---

1) Excluding the Ruwenzori and the high volcanoes the rainfall in various parts of Central Africa varies between 50 and 250 cm.

2) Presumably this is the case, since in the districts infected with the tsetse fly, where there is no cattle, and consequently the grass is not burned, trees do indeed grow.

An important result of this vegetation is that the sparsely planted high grass is unable to retain any fine detritus, certainly not on hill slopes and in fact not even on gently sloping plains.

With the sudden and intermittent rains only very little water penetrates the hard and dry ground, most of it flowing away over the surface and carrying off all the fine detritus.

The sub-arid climate is best developed when there is just enough rain to leave no places devoid of vegetation and the wind element has no part in the sculpturing of the landscape.

Photo 1. View from road leading from Mbara to Masaka.

As Davis has so clearly explained (Journal of Geology, Jan. 1930), the great difference between this and the humid cycle is that in regions where the latter prevails a dense covering of vegetation, particularly at the foot of the hills, prevents the fine detritus from being washed away. Evidence that the influence of vegetation is not over-estimated is furnished in a forested humid region where too many trees have been felled or destroyed by fire, for then enormous volumes of loose earth are washed away by the rain and it takes a long time for the vegetation to regain its former density. Furthermore in a humid region the ground, which is always more or less moist, easily absorbs all the rain water, which therefore does not displace the fine detritus by flowing rapidly over the surface, and this explains the concave

Photo Oosterchrist.
Photo 2. View from Kaina of arena wall facing east.

hyperbolic shape of the profile of a mountain.

A scarp in a sub-arid climate, on the other hand, is gradually broken up by physical weathering [1]) and in the beginning there is a great deal of waste. After the scarp has been worn down to a certain angle of slope, however, denudation is just as intensive near the foot as it is near the top and the angle of slope will not continue to decrease to the same extent if only the waste at the bottom of the slope is carried away.

---

1) It is not to be forgotten that the sub-arid climate is a continental climate and that there is a very great difference between the temperature in the sun in the daytime and that at night; in Ankole about 120° F. and 55° F. respectively.

Since all the hill slopes in Uganda have practically a constant angle from the base to the top (cf. photo 2) the waste must, therefore, be carried away, and at first the author could not understand how this was possible, seeing that there

are no tributaries on the wide and gradually sloping plains extending from the foot of the hills to the river; there are, it is true, some shallow gullies (photos 2 and 3), but these hardly ever contain running water and even in the rainy season only form stagnant pools.

Photo Oosterchrist.

Photo 3. View of arena wall near Kaina facing west, showing quartz reef dying out on Nobugumba hill in middle of photo; on plain to the right Kaina House; on hills to left waste from tunnel visible.

The slope of these plains varies, but on an average it is between 4 and 6 degrees. Plain and hill slope meet each other at a strikingly obtuse angle, without, generally speaking, any intermediate angle of slope. This is clearly seen on map 4, and even better on the three sheets of the map of the Anglo-German Boundary Commission. Obviously the climatic conditions favour two types of slope, namely a steep one of about 20 degrees, more or

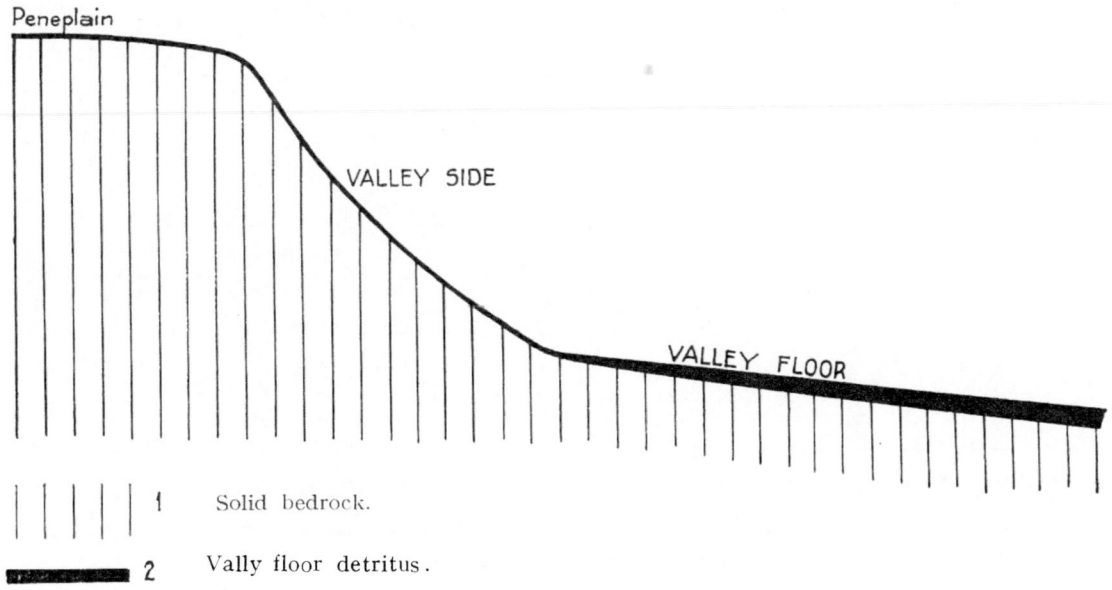

Fig. 1 Schematic section of valley floor and valley side.

less abruptly changing into one of about 5 degrees. It is evident that the plain is formed by the recession of the steep slope. In this paper the plain will be called the valley floor.

The waste from the hill side must, therefore, be transported in some way or other over the plain to the river, and Behrend believes this is done by "flood sheet" ("Flächenspülung"), which he describes as follows:

"Da aber die starken Wolkenbrüche der Regenzeit eine ausserordentlich "starke transportierende Wirkung haben, so werden die Verwitterungsprodukte "in ausgedehntem Masse weithin verfrachtet, und da die von den Gebirgen "niedergehenden Wassermassen sich oft in der Tiefebene kein dauerndes Bett "graben, sondern bei ihrem plötzlichen Auftreten die Ebene flächenhaft über- "schwemmen, so werden die von ihnen mitgerissenen Schuttmassen auch "flächenhaft verteilt, die Wolkenbrüche bewirken Flächenspülung . . . ."

Nowhere in Ankole has the author found evidence of coarse detritus being transported by flood sheet. Consequently, assuming that only the fine detritus is carried away, the hills would gradually sink, as it were, in the accumulation of coarse detritus at the foot.

Pit-digging has conclusively proved that the detritus does not increase in thickness towards the foot of the hill (see fig. 1), and although a view of the landscape sometimes leads one to suppose otherwise, it may be taken for granted that at the foot of the hill there is only a thin layer of detritus.

As will be explained in the following chapter, the writer has found that the detritus is transported both by flood sheet and by creep.

CHAPTER 2

THE VALLEY FLOOR, CREEP AND FLOOD SHEET

At the northern foot of the Ihunga are the head-waters of the Chabakaz (cf. map 4), lying in a basin of granite closed in on the south by the Ihunga and on the west and east by flat ridges. These ridges are likewise composed of granite, as also the lower slope of the Ihunga, the upper slope consisting of sediments (cf. section 4, Part II). On the northern side of the basin the Cha- bakaz flows through a narrow transversal valley cutting through the hill ridge of the Ruamheji — Nya Mirima (cf. photo 4).

This basin is formed by a number of gently sloping ridges in radial formation with shallow valleys in between, in which, however, there is no running water. At various places coarse cassiterite was found in the grass and this led to a number of pits being dug, but none of these yielded any coarse cassiterite. By chance, however, in the bottoms of the pits three pegmatite veins were discovered, intrusive in the granitic bedrock, and these indeed contained coarse cassiterite.

A fresh show of coarse cassiterite on the surface was then systematically

Photo 4. View from north slope of Ihunga facing headwaters of Chabakaz and Ruamheji Ridge; gorge of transverse valley of Chabakaz hidden behind grass roof to right.

explored (see fig. 2). The grains of the cassiterite were ½—1 ½ cm in diameter and fairly well rounded at the edges. The spot where it was found was some 5 metres in diameter and the top soil contained as much as 1 % of cassiterite.

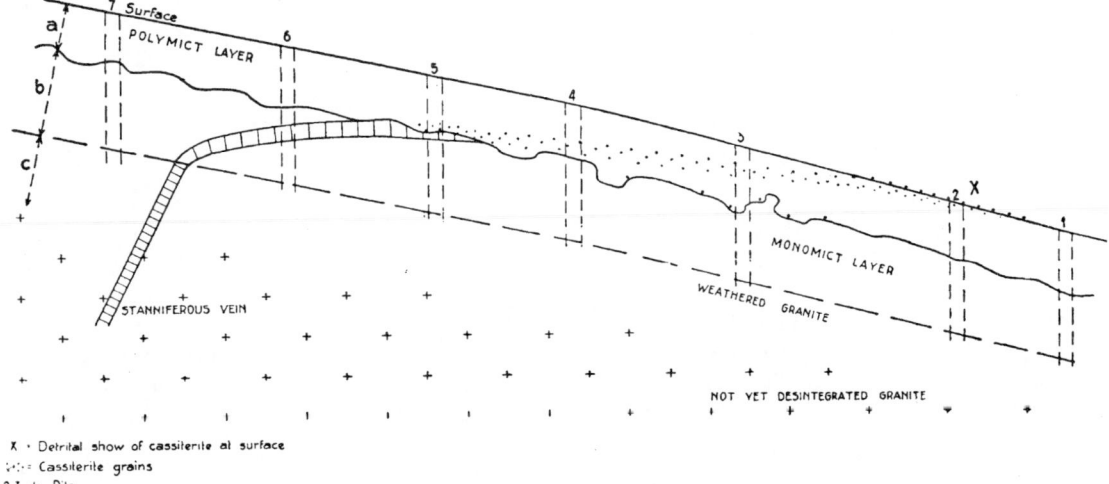

Fig. 2. Section of valley floor observed in a trench on the north flank of the Ihunga.

Pits dug higher up the slope disclosed cassiterite in the 3—4 ft thick top detritus layer, which was characterised by the presence of granitic components (quartz and feldspar) as well as sedimentary material (polymict layer = layer *a* in the accompanying illustrations). Among the sedimentary materials the most striking were fragments of garnet quartzite. In pits dug still higher up the slope this detritus layer no longer contained coarse cassiterite, but this mineral was found in the next lowest layer of detritus, which was characterised by the absence of sedimentary material (monomict layer = layer *b* in the illustrations). The writer ascertained that this latter layer could not yet be called a normal granitic bedrock, since all parts were displaced in respect to each other; this became evident when digging a deep trench through the pits extending into the only slightly weathered granite.

In this trench a pegmatite vein was encountered containing cassiterite, and in the weathered granite (shown as *c* in the illustrations) this vein was a little broken. Higher up, however, the vein turned valley-wards and in the monomict

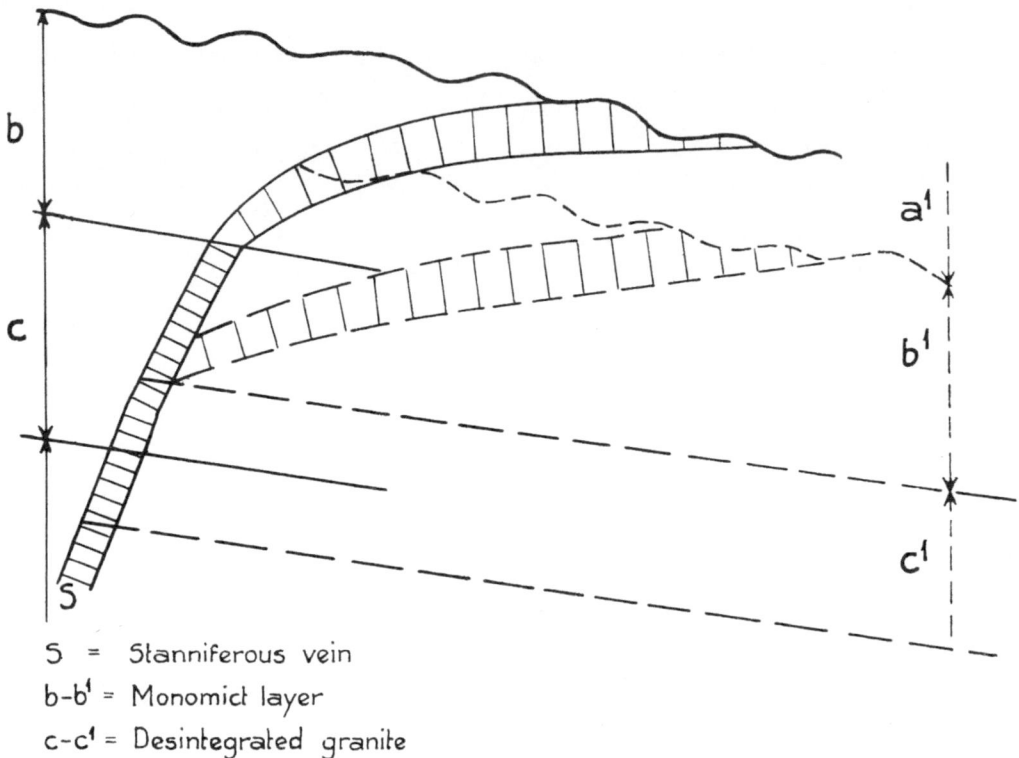

S = Stanniferous vein
b-b¹ = Monomict layer
c-c¹ = Desintegrated granite

Fig. 3. The lowering of the bedrock near a stanniferous vein in course of time. Continuous lines indicate the situation at a given moment. Broken lines indicate the situation after the lapse of some time, during which disintegration of the granite has advanced in depth (c—c¹); creep has affected deeper levels (b—b¹) and part of the monomict layer has been incorporated in the polymict layer (a¹).

layer the broken fragments became more and more scattered, until near the less defined, undulating interface against the upper polomict layer the vein ceased. The whole had the appearance, therefore, of the tail of a comet. The monomict layer thus consisted of granitic detritus, the particles of which had been displaced in relation to each other by creep.

Particles of the polymict layer (a) overlying the monomict layer have already advanced several kilometres, whilst the parts of the monomict layer (b) have been displaced some ten metres at most.

The latter, monomict, layer is underlain by disintegrated granite (c), which in turn changes into weathered but not yet disintegrated granite.

It is over this compact granite that the rain water penetrating the earth gradually flows down to the river. This ground water further weathers the granite and saturates a more or less thick layer of the detritus, causing the latter to creep; the manner of this process will be described farther on.

The polymict layer (a) creeping over the monomict layer (b) becomes mixed at the bottom with particles of the latter layer and is continuously withdrawing material from it (see fig. 3). Consequently not only the top of the solid granite but also the top of the monomict layer is continuously falling. This interface between monomict and polymict layer is commonly called "bedrock", but it is obvious that the boundary of the bedrock is very difficult to define.

If there were no other factor than creep, layer *a* would gradually incorporate cassiterite grains from the vein and carry them along towards the river, but it is seen that the cassiterite grains reach the surface fairly quickly and almost

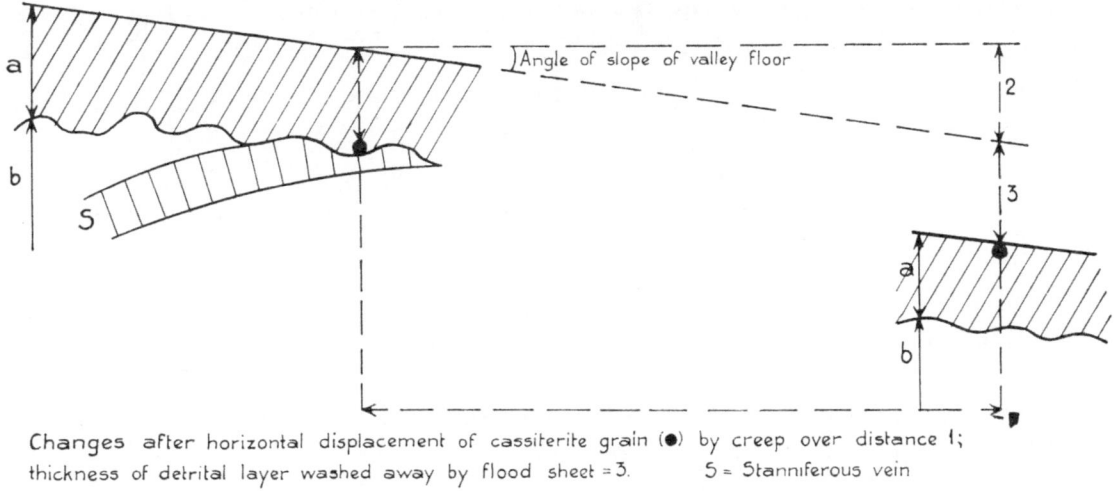

Changes after horizontal displacement of cassiterite grain (●) by creep over distance 1; thickness of detrital layer washed away by flood sheet = 3. S = Stanniferous vein

Fig. 4. a = polymict layer, b = monomict layer.

simultaneously, for a short distance valley-wards from the place where the cassiterite was found layer *a* no longer contained any coarse grains of this mineral. This is due to flood sheet. A shower of rain covers the surface with a thin sheet of water which, flowing rapidly down towards the river, takes with it the fine and specifically light detritus — which moreover is also carried towards the river by creep and after every shower is transported a little distance farther — whilst the coarse or specifically heavy fragments are left behind. In this manner the fine material reaches the river much quicker.

Thus the cassiterite comes to the surface in consequence of the covering layer being washed away, while the grains gradually creep towards the valley (see fig. 4); cassiterite from veins which have no outcrop becomes concentrated on the surface (the top soil contained 0.75 % Sn, whereas none of the veins contained more than 0.06 %).

Consequently in course of time not only the interface between layers *a* and *b* but also the top of layer *a* falls (see fig. 4). Deductions made from the data available showed that where the polymict layer had moved 60 metres, with a slope of 6 degrees, the surface had dropped about 120 cm. Over a distance of 3 km from hill slope to river the landscape would therefore drop some 60 metres in the time taken for a coarse fragment to creep down from the hill slope to the river, assuming the same rate of denudation.

It is to be expected that at a short distance beyond the contact with the granite the polymict layer will already be found to consist for the greater part of granitic components, and that all coarse or specifically heavy fragments will be concentrated at the surface. And this is indeed what happens, though incompletely, for in the whirling movement set up in the interface between the polymict and monomict layers fragments of quartzite and grains of cassiterite are held back.

The general tendency, however, is for the coarse fragments to accumulate at the surface, while valley-wards the polymict layer is found to consist almost entirely of granite fragments.

The fact that the specifically heavy minerals come to the surface has greatly facilitated the exploration of the valley floor and the remnants of the peneplain on the hill ridges (which latter, naturally, undergo the same processes as the valley floor), as the absence of outcrops of stanniferous veins through the thick detritus layer made exploration extremely difficult. On the other hand, however, it was apt to be misleading, since cassiterite found on the surface may have come from a great distance and its presence is not to be taken as an indication of a particularly rich detritus.

The cause of the creep is to be sought in the action of the ground water, in combination, of course, with gravity.

The influence of the ground water was deduced from the distribution of extremely fine cassiterite, which in most cases was the only product obtained from the numerous pits dug near the Ihunga. This metal was disclosed by analysis of the fine concentrates, which consist chiefly of iron oxides. As a rule the polymict detritus layer contained per cubic metre about ½—1 gramme of cassiterite in this form, but in the dry valleys, which undoubtedly function as drainage channels for the groundwater, the content was 5—6 grammes, whilst in pegmatite veins as much as 8 grammes per cubic metre was found [1]). The distribution and concentration of this fine cassiterite must be ascribed to the action of the ground water in transporting the smallest particles of the detritus along minute channels. The resultant undermining, so to speak, of the large grains must lead to creep; these large grains tilt over and under the influence of gravity the whole mass of detritus moves towards the river.

The presence of coarse cassiterite on the surface further proves that, when undermined and tilted, the cassiterite, in spite of its comparatively great specific weight, does not sink, at least not so quickly as the surface is caused to fall by flood sheet.

---

[1]) The fine tin found in the polymict layer may partly have originated from the granite.

Having thus explained the process of transportation over the valley floor, the author wishes to digress for a moment to deal with the effect of river degradation and aggradation upon the development of the valley floor.

When the transport capacity of a river is greater than the volume of detritus brought down to it, the river will deepen its bed. If this is done slowly, so that no cañon is formed (and thus no new scarp), the velocity of the creep will increase, particularly close to the river, and this increase of velocity will gradually advance to higher zones of the valley floor.

The weathering of the valley sides, however, is not increased. Thus with the same rate of addition of detritus but with increased creep velocity the polymict layer will decrease in thickness. On the other hand, with higher creep velocity the polymict layer will take up more material from the monomict layer, and the interface between the two will fall more rapidly than before. The mainly chemical weathering of the granite, however, cannot be hastened. The extreme case is when the polymict layer immediately overlies a granite that has been but slightly weathered, and this is more likely to occur in the higher zones of the valley floor than in the lower zones; close to the river, where weathering has already taken place for some time and ground water is always present, the granite will be more deeply weathered than in the higher parts of the valley floor, where as

Photo 5. View of Rushenyi arena from garden of Kaina House facing towards the Chamiombu, highest mountain in middle of photo. Hummock of bare granite is seen in the undulating plain to the right.

a matter of fact bare hummocks of granite protrude through the detritus (see fig. 1 part II, Ndorwa arena, and photo 5, Rushenyi arena). Consequently degradation of the valley floor in the higher zones soon becomes very difficult, whilst close to the river where the weathered layer (indicated by c in the illustrations) is of great thickness, degradation is able to advance for a much longer period.

As a result the angle of slope of the arena floor will increase and with it also the effect of flood sheet. The fine material will be transported much quicker and finally the polymict layer, consisting chiefly of coarse fragments, will come to lie on a slightly weathered bedrock, while in the meantime the dip of the valley floor will have increased by a few degrees. The existence of a "bare rock pediment", which Davis takes to be possible, does not appear to the author to be at all likely, for the coarse and specifically heavy parts can only be transported to the river by creep.

Should, however, the transport capacity of the river diminish, for instance through the slope of the surface being reduced in consequence of tilting, then there will be an accumulation of detritus near the river, and the polymict layer will creep more slowly, increase in thickness and take up less material from the monomict layer. The weathering of the granite, however, will continue as before. Ultimately both the polymict layer and the monomict layer will become very thick and the hard and slightly weathered bedrock will come to lie at a considerable depth, whilst the angle of slope of the valley floor will have been reduced (see fig. 5).

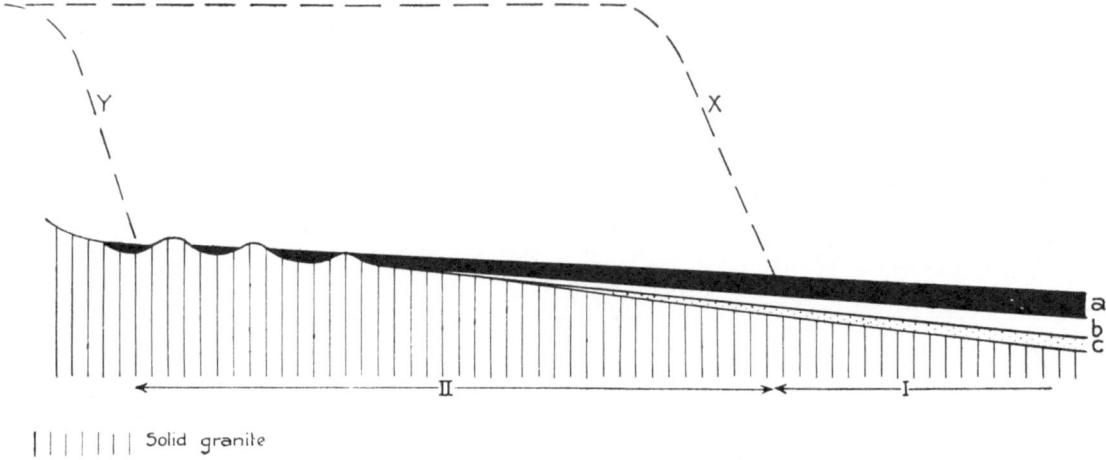

| | | | | | | Solid granite

Fig. 5. Section of valley floor showing hummocks of bare solid granite near the actual valley side and increasing depth of weathering away from valley side.

    a. polymict layer.
    b. monomict layer.
    c. granite weathered but not yet disintegrated.
  Part I longer exposed than Part II, latter only after recession of valley side from X to Y.

Although the extreme case of degradation does not occur, the basin of the Chabakaz river (which is tributary to the young Kahinji) shows the phenomena of degradation just described, particularly in comparison with Rushenyi and the Kyerwa arena, where the detritus is very thick indeed.

CHAPTER 3

VALLEY SIDES — FORMATION OF ARENAS

The valley side rises precipitously from the valley floor with a practically constant angle of slope from the base to the top. Such a slope is only sparsely grown with grass. The detritus layer lying on the bedrock (this can be compared to the polymict layer *a* described in the previous chapter) consists principally of coarse fragments and is no more than 6—12 inches in thickness. The upper part of the bedrock is weathered and broken and the strata have already been bent valley-wards (thus the equivalent of the monomict layer *b* in the previous chapter).

The rapid changes in temperature, the rains and the roots of grass all promote weathering. The coarse fragments gradually roll down the valley side and the fine materal is immediately washed away.

There being no dense growth of bushes and trees to retain the fine detritus at the bottom of the slope, there is no gradual reduction of the angle of slope.

Consequently steep scarps are not confined to any stage of the sub-arid cycle, whether young or mature.

Davis has made this quite clear, and to quote from him:

"A striking difference between humid and arid mountains is found in the "decreasing angle of slope in the first as the erosion cycle advances, in contrast "to the constant angle of slope „in the second".

The conclusion is, therefore, that the sub-arid climate gives rise to a special kind of vegetation and in consequence the erosion process differs considerably from that of the humid cycle.

Since in Ankole the valley floor is often found to be composed of granite and the valley sides mostly of sediments, it would be obvious to suppose that the steepness and constant angle of slope of the valley side are subsequent results of a difference in the composition of the bedrock. It might also be supposed that the discontinuity at the foot of the slope is due to the same cause. Both these assumptions, however, are incorrect.

Photo 6. Katuba valley with river to the left, showing contrast between valley floor and valley side; habitations mainly at foot of latter, the only place where running water is obtainable.

At Kaina (section b, Part III) this discontinuity is quite noticeable, notwithstanding that a large part of the mountain side and also the valley plain consist of granite (cf. photos 2, 3 and 7). The same is the case with the slopes of the Chamiombu and Nerionza. At the foot of the Nyihanga the discontinuity lies in sediments, of which material also the whole mountain side and the adjacent part of the plain are formed.

The valley of the Nya Buganda, east of Lutobo, though formed entirely of granite still shows discontinuity in valley side and valley floor, as is also the case in the Katuba valley, though the latter is formed entirely of sediments (see photo 6).

The Kivimbiri valley (east of Ruberogoto — map 3) likewise has a floor with a clearly discontinuous transition to the valley sides, though here again both floor and sides consist of sediments.

Another explanation that might be advanced is that the scarps were due to the presence of some hard rock, since on almost all mountain slopes there is an outcrop of a quartzite or quartz vein. But this, too, is incorrect. Both on the north and on the south side of the Ruamheji (photo 4) or the Kavusanam-mi-West there is an absence of relatively hard rocks and yet the scarp rises suddenly. The same is to be seen on the west side of the Chamter. A decidedly hard rock is not the cause of the formation of a scarp, as may be seen, for the matter of that, in the section of a mountain slope (fig. 6): the steepness of the

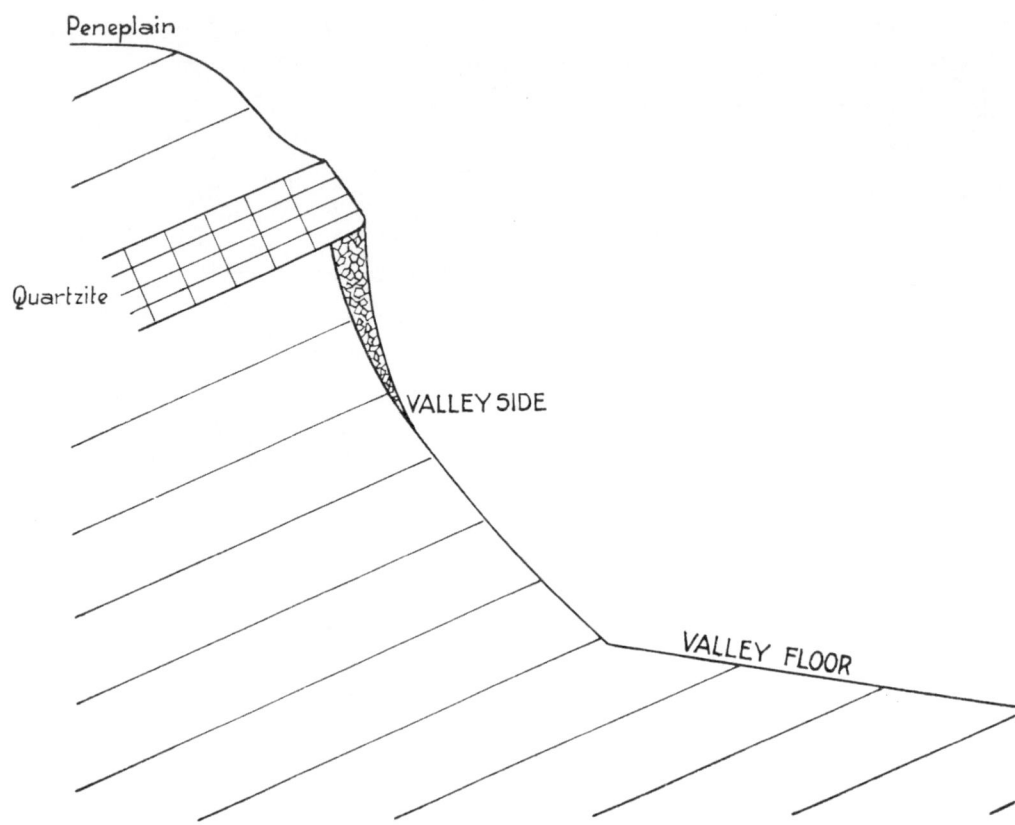

Fig. 6. Effect (exaggerated) of resistant rock outcropping on valley side.

scarp is regular both above the hard layer and below it. It is only imme-diately underneath a quartzite or quartz vein that the scarp is more precipitous (cf. photos 2 and 7): here it crumbles away along joints and bedding planes and is more or less undermined. The scarp extends, however, from the valley floor to the peneplain covering the ridges and hill tops.

In the middle of Uganda, near Kampala (photo 8) and Masaka, for instance, this peneplain is very clearly seen everywhere; the hill ridges are perfectly flat and horizontal. Farther south the peneplain changes into a gently undulating hill country with monadnocks. Nevertheless it is clearly distinguished on the top of the mountains (cf. map 4, SW. corner), while for instance on the long hill ridge encircling the Lutobo valley (cf. photo 7) the differences in elevation are only a few hundred feet.

It is supposed that Central Africa was warped in the Cretaceous, the peneplain disappearing to the east below sealevel and in the west dipping towards the Congo basin. Around Lake Victoria the elevation of this pre-Cretaceous peneplain is roughly 1300 metres and in some parts of Kenya about 3000 metres, whilst in SE. Tanganyika its cretaceous covering is being carried away by erosion.

In consequence of this uplift the rivers quickly cut out their beds (cf. photo 9). The scarps, therefore, are receded cretaceous cañon walls, which in the sub-arid climate (continuous since the Karoo) have receded with a constant angle of

Photo 7. Arena wall south of Kaina House. Remnant of peneplain clearly visible on ridge separating Rushenyi granite from Lutobo granite to south.

slope. In the sediments this recession took place more slowly than in the rapidly disintegrating granite, particularly because of quartzites and quartz veins forming a great though not insurmountable obstacle.

Photo 8. View from Kampala facing Mengo (King's hill). Remnants of peneplain in the background to the right.

Thus the combination of scarp and valley floor is the consequent result of a difference in height in a sub-arid climate, whereas in a humid climate this is expressed by the hyperbolic nature of the slope.

The areas formed by granite now appear as low undulating plains surrounded by a belt of steep mountains, the whole fully deserving the name of arena, in which the arena wall and the arena floor can both be distinguished.

The scarps have entirely disappeared in the granite, so that the ridges between the rivers are formed only by valley floors (see fig. 7).

These valley floors are first formed when large volumes of coarse waste roll down the steep cañon walls, thus forming on either side of the river a slope of waste which is no longer reached by the river. Later on, when the cañon wall has become a valley side and regularly recedes, the angle of slope of this embryo

valley floor decreases and the constituent materials become finer-grained. The
further stages are well depicted in fig. 8, representing a part of the Chitwe arena.
The contour lines in this sketch are drawn at intervals of 30 metres (100 ft)
and the geological data have been taken from Combe's and Miss Ledeboer's
maps. The arena wall is most pronounced in the north, where the contours
show the steep mountain slopes,
on the north side of which are
traces of the ancient peneplain.
It can easily be seen how the
peneplain between the two
rivers Lutembi and Zerukuro
has been worn away by the
recession of the valley sides,
the water divide between those
two rivers being formed by
gently sloping valley floors. The
discontinuity between the valley side and the valley floor is
also striking; on the valley floor
the distance between the contours remains practically con-

Photo de Greve

Photo 9. View of steep and narrow valley in Kigezi near
Kumba facing west. Remnants of peneplain visible. Kraal
of Chief of Kumba in foreground.

stant until at the foot of the valley side it suddenly decreases. While in the
granite area all traces of valley sides and peneplain have disappeared, they

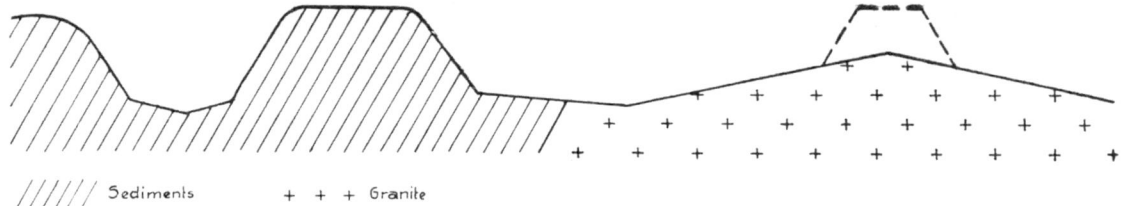

////// Sediments   + + + Granite

Fig. 7. Formation of arenas by slow recession of valley sides and sediments, and quick recession in granite.

are still in evidence over a considerable part of the sedimentary area. However, also in the latter area, and especially where quartzites are absent, the
peneplain between two rivers may have been worn away. The absence of
scarps between the rivers Lutembi and Zerukuro is undoubtedly due to the
absence of quartzites, whereas the recession of the valley side on the north
of the Chaingan river has apparently been retarded by the presence of the
boundary quartzite. Arena walls, generally, are formed by such retarded
valley sides, on which very often hard and resistant rocks are exposed.

Consequently the arena floor consists of an undulating sequence of valley
floors (cf. photos 3, 4 and 5). The peneplain is not entirely flat, the difference
in altitude between A and B being 90 metres, over a distance of 5 kilometres;
3½ km west of B is an old hill, probably a Monadnock (the boundary quartz-

ite?), 120 metres higher than B (hill C, altitude 6066 ft). The valley sides, however, have generally a height of 240 metres.

Another good example of an arena floor is given in photo 5, a view of the Chamiombu and the Shonobutondo taken from Kaina. The very small angle of slope and particularly the very regular incline of the floor of the Rushenyi

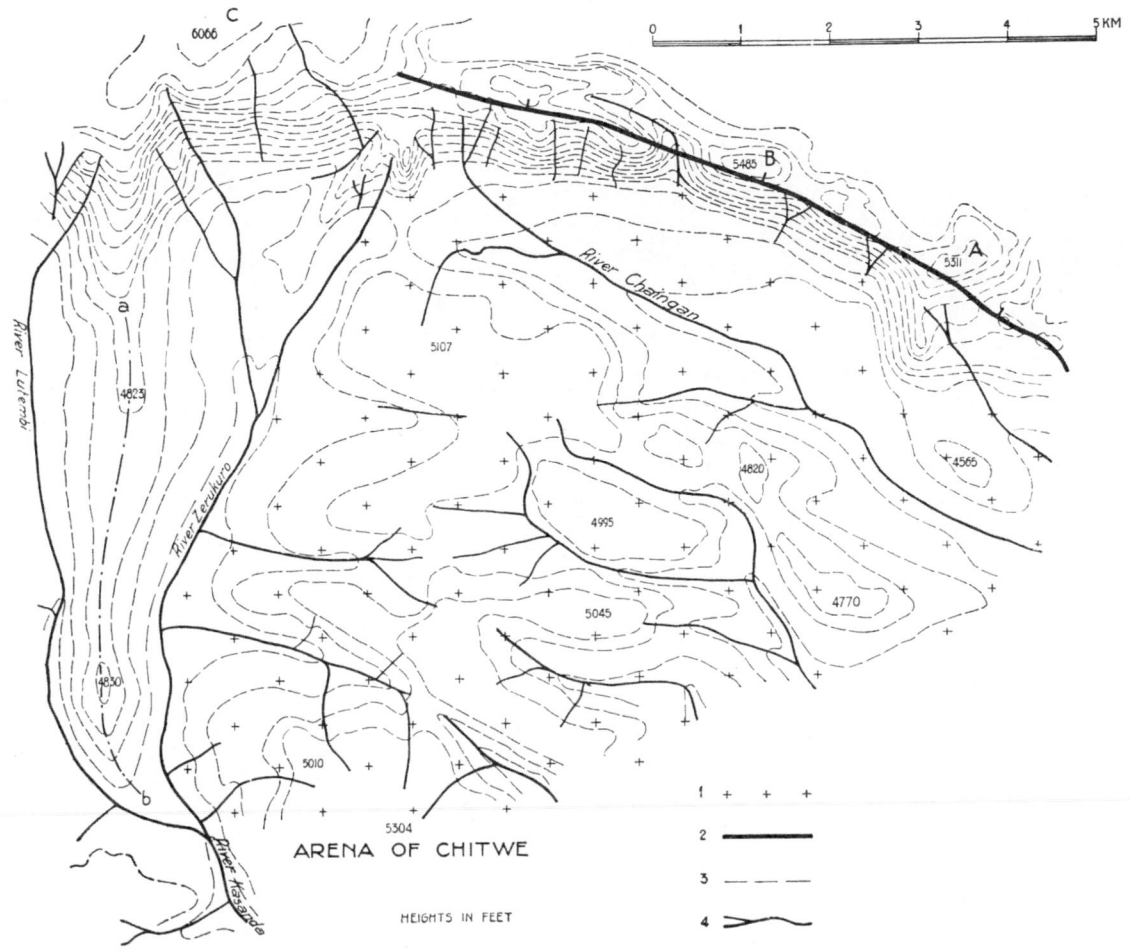

Fig. 8. Chitwe Arena.

1. Granite.
2. Boundary quartzite.
3. Contours (interval 100 ft).
4. Rivers, gullies or dry valleys.
A, B and C Hills in peneplain.

Ridge between rivers Lutembi and Zerukuro is formed by valley floors along A—B, valley sides having disappeared.

arena are clearly seen, as also the undulation of the arena and the succession of valley floors. The first ridge seen in the arena shows granite boulders and hummocks of bare granite near the arena wall. In front of the ridge is a dry valley such as has been described. The gentle slope of the arena floor contrasts sharply with the steep hill ridges rising in the distance. Photos 2, 3 and 7 show the arena wall at Kaina; discontinuity is fairly marked here.

Photo 10 was taken from Lugalama in the middle of the arena facing west, with the Ihunga in the background. The Lugalama arena in the foreground is bounded on the west by a steep ridge of hills pushed forward in the manner of a piece of side scenery between the Ihunga and the arena and broken in the middle by a wind-gap (altitude 5500 ft). Presumably this arena has been captured several times.

Photo 10. View from Lugalama facing west to the Ihunga in background. The foreground is undulating arena floor, with arena walls to left and right of the Ihunga.

As the valley side recedes, so its height, measured from foot to top, diminishes, until finally there is but a small isolated remnant of the peneplain left. It is in this manner that "Inselberge" are formed (cf. fig. 9), but it seems to be beyond question that the same cause does not apply to the formation of residual hills in Ankole, since owing to the varied composition of the sub-surface monadnocks are more likely to be formed. "Inselberge" are the purely

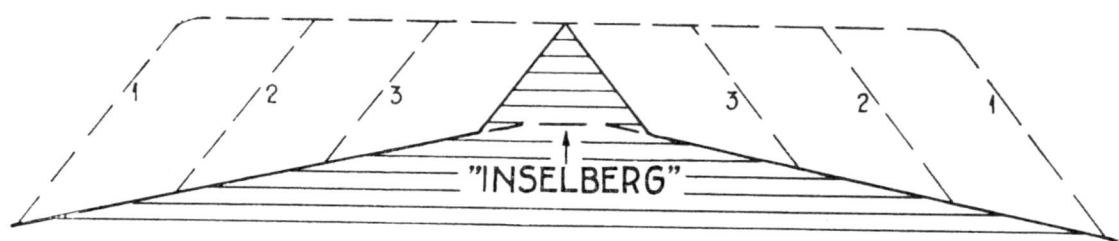

1-1, 2-2, etc = valley sides in course of recession

Fig. 9. Formation of "Inselberge" through recession of valley sides.

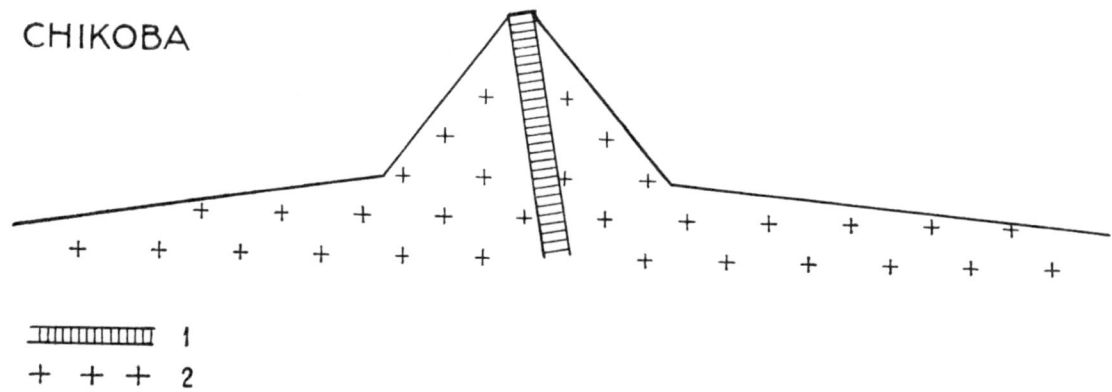

Fig. 10. Nature of isolated mountains in granite area of Wishikatwa.
1 = roof pendant consisting of quartzite enveloped or not by remnants of schist.
2 = granite.
The Chikoba is not merely a consequent residual hill but a monadnock, the quartzite retarding denudation.

consequent remnants typifying the gneiss and granite plains of Tanganyika, Kenya, Mozambique and Nigeria, where the composition of the bedrock is markedly homogeneous.

There are, it is true, also isolated mountains in Ankole, but their subsequent origin is obvious. As an example the Chikabo (cf. maps 4 and 5) may be taken (fig. 10): the eastern and western valley sides have approached each other and the peneplain has disappeared, the situation of the mountain now being fixed for ever to the course of the quartzite, which forms its backbone. The subsequent character is all too evident than that such mountains could ever be called "Inselberge"; they are simply "monadnocks".

CHAPTER 4

CONCENTRATING POWER OF THE RIVERS

Every river valley began as a cañon after the upheaval referred to in the previous chapter. At that time water flowed through, and in many of the valleys which have now become dry, coarse boulders are still to be seen in the detritus (on the monomict detrital layer).

As the valley floors spread out, the water loses itself more and more in the detritus and in course of time there will no longer be any running water over the detritus, not even in the wet season. It is only in the ultimate river, such as the Rufua from the Rushenyi arena, the Kachwamba from the Lugalama arena and the Kagera from the Kyerwa and Ruberogoto arena, that water is still running over the detritus.

The branch valleys, on the other hand, only contain ground water. This promotes creep, and through creep and flood sheet the layer of detritus (in these valleys very thick) is transported to the ultimate river. Consequently there is no concentration of specifically heavy minerals in these branch valleys. Figs. 11 and 12 show how transportation takes place in an arena under humid and under sub-arid climatic conditions respectively.

The conclusion to be drawn is that alluvial tin deposits can only be formed in the ultimate rivers, with flowing water in their beds.

Wayland has another explanation for the absence of alluvial deposits in such gullies. He believes to have proved with the aid of prehistoric objects that ice periods in Europe corresponded to pluvial times in Africa, in which times game, and, following game, mankind penetrated Central Africa. It is to these pluvial conditions that he ascribes the cause of the absence of alluvial deposits in the small valleys that are now mostly dry, the deposits having been washed away. That this explanation cannot be accepted will be evident from the following.

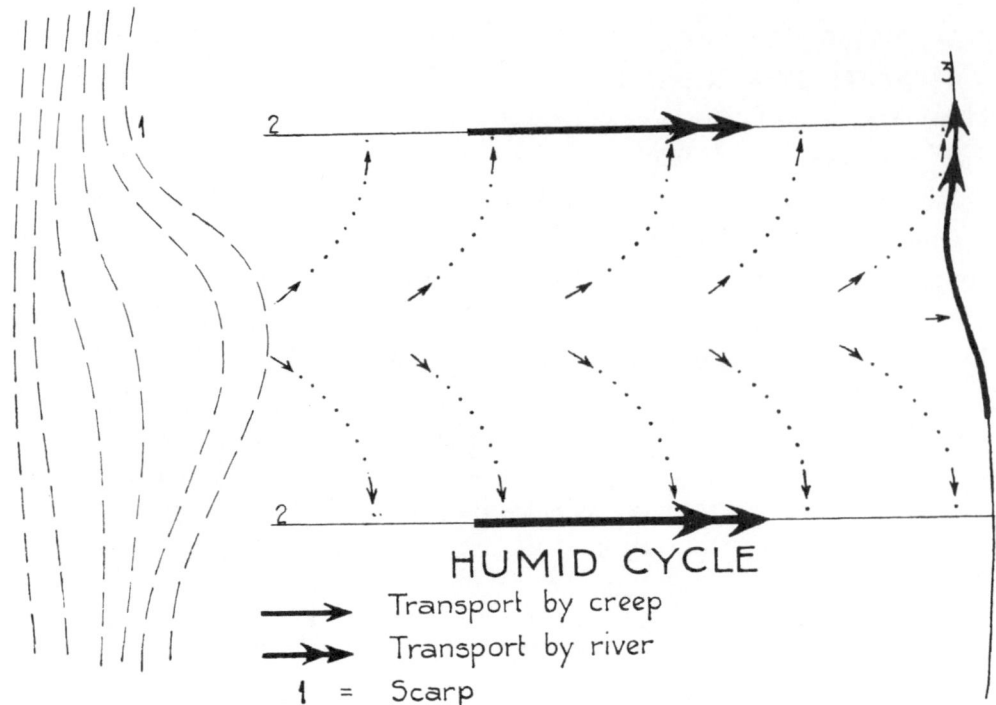

Fig. 11. Conditions in humid cycle; 1 — scarp, 2 — tributary, 3 — main river. River transport being much quicker than creep, nearly all the detritus reaches the main river via the bed of the tributary. Length of arrows indicates relative velocity of transport.

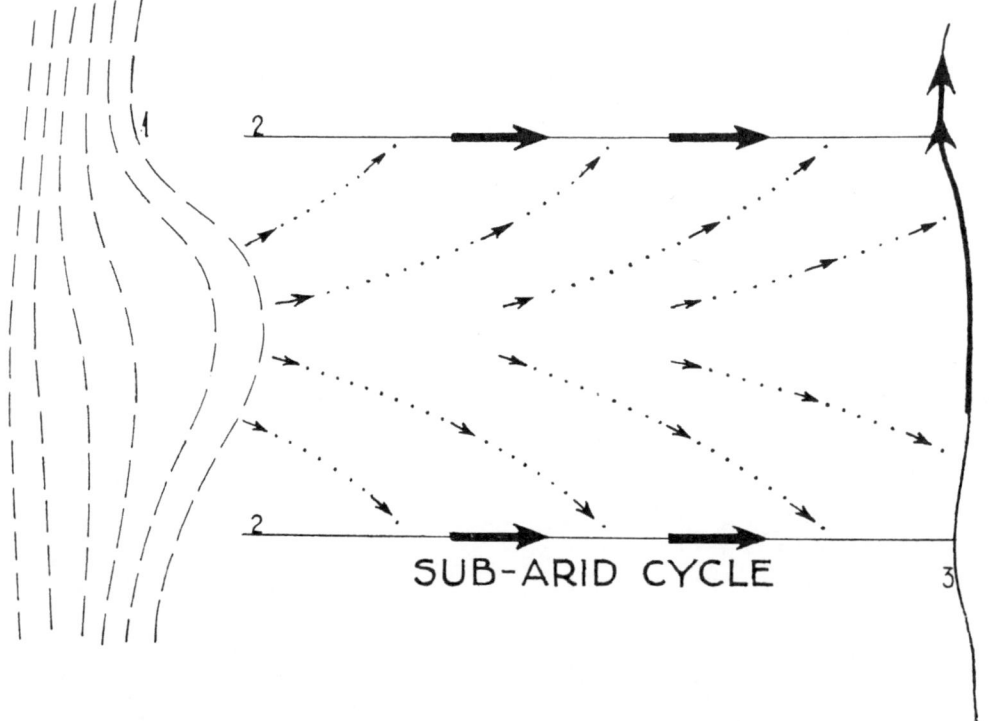

Fig. 12. Conditions in sub-arid cycle; 2 — dry valleys, 3 — main draining river. Transport in dry valleys is by creep, the velocity of which is only slightly greater than that on the interfluvial ridge, owing to abundance of ground water. This leads to the detritus on the middle of the ridge gliding more directly to the main river.

Considering that also at the present moment game and mankind are abundant in Central Africa, it is quite possible that present conditions in those parts are quite similar to these prevailing in the pluvial period. Nevertheless it is a known fact that even with a rainfall sufficient for the existence of mankind and game in large numbers the majority of the valleys never contain running water, thus precluding the possibility of heavy minerals being washed away.

Moreover, even if it were assumed that during those former pluvial periods the total annual precipitation was much greater than it is now, we are still absolutely in the dark as to whether these conditions actually constituted a humid cycle, and it is only in a humid cycle that an ultimate degradation of the valleys is possible, so that deposits of heavy minerals may be washed away.

Finally, if a humid cycle had prevailed its effects would anyhow have been neutralized by the effect of the tilting of Uganda, which in consequence of the Rift Valley movements had already taken place.

In the case of the westward-flowing river system this tilting caused the slope of the rivers to be reduced and in some cases the direction of flow was even reversed. Consequently the transporting power of the rivers was reduced to nihil and many valleys were transformed into lakes, as may clearly be seen on map 1. In Ankole, Ruanda and Karagwe whole arenas were transformed into complicated swamp systems, a good example of which is still to be seen in the Kyerwa arena (cf. map 3).

Nature's remedy is aggradation; the valleys collect detritus, the river bed is raised and finally the transporting power of the river becomes equal to the volume of waste added.

In such rivers, then, even a large influx of water cannot have led to any renewed erosion and removal of alluvial deposits,

Both in the Kyerwa and in the Rushenyi arenas river erosion has been at a standstill during a long period; the consequences have already been mentioned. *If* pre-miocene or pre-oligocene alluvial tin deposits have been present in the Kagera they must still be lying in the old bed buried somewhere in the enormous swamps underneath detritus and bog. In consequence of the Rift Valley movements, however, some rivers or parts of them were rejuvenated, as is the case with the Kahinji, to which the aforementioned Chabakaz is tributary (cf. maps 1 and 4).

The difference between the Chabakaz basin and the arenas of the middle Kagera is clearly shown by the thickness of the detritus in the dry branch valleys, north of the Ihunga the thickness being only some 20 ft., whereas in the Kyerwa arena the top of the bedrock in some places lies as much as 80 ft. below the surface.

The alluvial deposits in rejuvenated rivers were undoubtedly removed, not necessarily by the heavy precipitation of pluvial times, but rather by an increase of the gradient of the river bed in a sub-arid climate.

CHAPTER 5

EVIDENCES OF WARPING — SWAMP DIVIDE

Map 1 clearly shows that the central part of Uganda is a drowned basin. The fantastic outline of Lake Victoria contrasts sharply with the regular shores of the Rift Valley lakes. In many places around the Victoria Nyanza the outlines of a drowned valley system are to be recognised, as for example the Sese islands in the lake just east of the estuary of the Katonga. Lake Kyoga clearly betrays the fact that really it is the drowned valley system of the Kafu. The trend of the creeks and of the tributaries gives the impression that both the Kafu and the Katonga flow westward, whereas in fact at the present time these rivers flow sluggishly eastward through broad, swampy valleys, rising some-

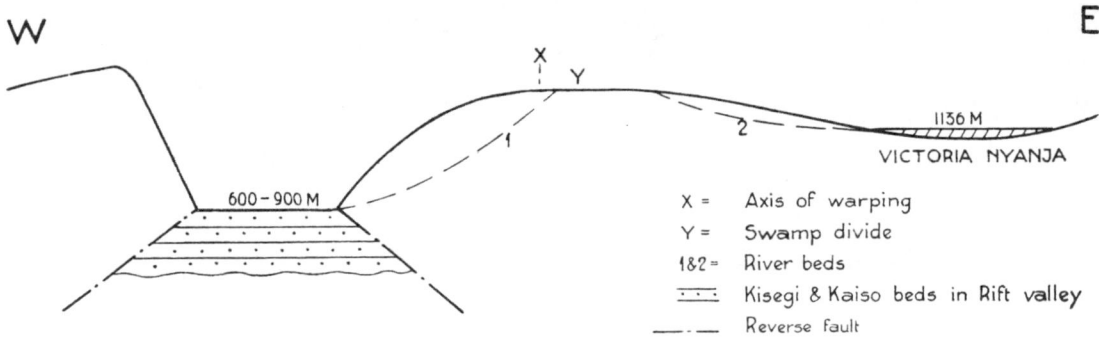

Fig. 13. Formation of Rift Valley according to Wayland's compressional hypothesis.

where in a broad valley. The same is the case with other rivers, e.g. the Rufua. They rise in a swamp or lake (e.g. Lake Karenge, maps 3 and 4), from which a young river flows rapidly westward through a deep valley towards the Rift Valley, while another river flows sluggishly eastward in a valley filled with reeds and papyrus. The walls of the valleys, however, continue undisturbed along the water divide, and this is called the swamp divide.

Wayland's conception of this phenomenon (see fig. 13) is that to the west of the basin there is a convex land area dipping very gently eastward and rather steeply westward. The latter flank becomes very steep and forms the eastern wall of the Rift Valley. Thus the erosion on this side is much greater than on the side towards Lake Victoria, since the distance from the swamp divide to the 600 or 900 metre level is much less in the westward direction than in the eastward direction via Victoria Nyanza, which is 1136 metres above sea level. Consequently the divide is gradually shifting eastward.

As a result of this warping the headwaters of most of the rivers have been drowned (cf. map 1), an example of which is the Kanyamagogo in Kigezi. This river, however, is a tributary of the Kahinji, which flows into Lake Edward. In its lower course the Kanyamogogo is a young stream, whereas the upper reaches lie in a broad swampy valley (photo 11). Owing to receding erosion the young stream gradually clears away the detritus and swamp in the bed and thus gives back to the latter its normal profile (fig. 14).

Photo 11. Drowned upper course of Kanyamagogo near Soko facing north-west.

The height of the culmination of the warped area, however, may exceed that of an arbitrary point of the original divide (see fig. 15), in which case the water will flow over and cut out a deep valley. This is in fact what happened with the Kirurama and the Emise, which formerly constituted one river running NW. After the warping the water no longer continued to flow over the culmination in NW. direction, but ran over the water divide eastward into the valley of the Upper Muvumba, which feeds Lake Victoria via the

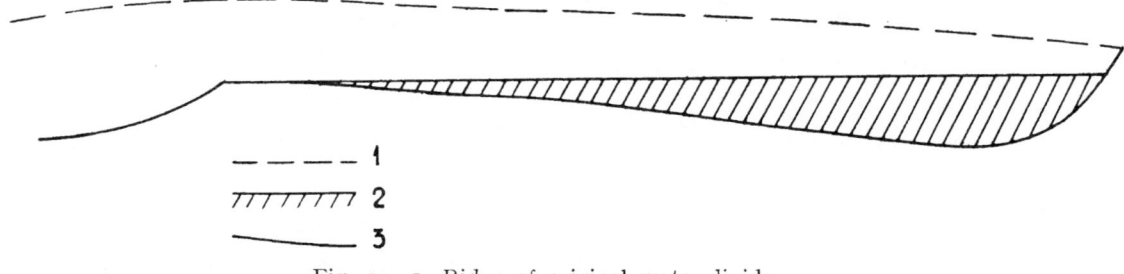

Fig. 14. 1. Ridge of original water-divide.
2. Surface of swamp.
3. Warped river-bed.

Kachwamba, Kakitumba and Kagera. Now the Kiruruma flows southward and the Emise northward, both feeding an ungraded and not yet adjusted stream rushing over small waterfalls through a narrow and deep valley.

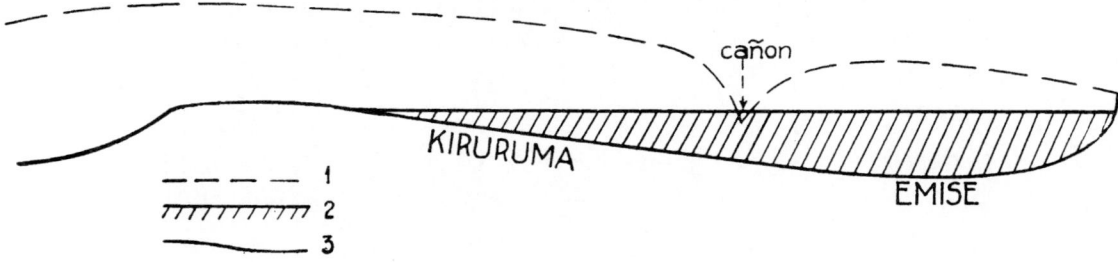

Fig. 15. 1. Ridge of original water-divide.
2. Surface of swamp.
3. Warped river-bed.

On the NW. side of the swamp divide are the headwaters of the Ishasha, which farther on likewise flows into a deep valley and near Kumba is now diverting the waters of the Kashungati from the Kiruruma river (cf. map 6).

A glance at map 1 will show that these swamp divides lie in one line, which from a point SW. of Lake Karenge runs practically over Nya Lusanje (map 3) towards Kumba. Bearing in mind that the swamp divides no longer indicate exactly the axis of the culmination but have already been displaced somewhat to the SE. or E., the axis of the culmination should lie somewhere near the Wramuyonyi and the Kashuli (see map 6); the first named mountain is in fact the highest point in this area.

It is impossible to map the continuation of the axis of the culmination farther to the south-east.

South-west of the Kayonza Forest there are other divides similar to those described above. At Lubuguli, for instance, the valley sides continue unbroken from the SE. to the NW., but on one side of the village the Kashash (farther downstream called the Irwi) flows NW, whilst on the other side there is the Nokaskarara flowing eastward (cf. map 6). Obviously the Kaferongo once flowed towards the NW. via Lubuguli. At present, however, the Nokaskarara and the Kaferongo are tributaries of the Ruezaminda, which flows through a transversal valley cutting through several SE-NW mountain ridges and empties its waters into Lake Mutanda; this lake lies in a branch of the Rift Valley. Rising up on the south side of the transversal fault-trough are the Mufumbiro volcanoes, the Muhavura, the Mgahinja and the Sabyinyo (cf. map 3). In course of time a huge quantity of lava has been poured out into this fault-trough and as a result of damming Lake Mutanda was formed between the lava and the mountain country built up by the Karagwe-Ankolian system.

Lake Mwanga and the Kafuga river likewise lie in one valley, which extends farther to the west, beyond the source of the Kafuga.

The question is how these divides were formed. A study of the geological history of the peneplain makes the answer quite clear. At some time or other the peneplain had a westerly dip, but owing to tilting there is not much trace of this left. When following the ridge from the Ruakwataro (2531 metres), via the Butengo (2320 m.) and the Kamena (2390 m.), to the Naiguru (2413 m.) one first finds traces of warping west of the Kasatora (2434 m.), from which point the land falls sharply to a level of about 900 metres in the Rift Valley near Rutchuru.

The same applies to the ridges to the SW. of that just mentioned.

Proceeding, however, from the first mentioned ridge in a southwesterly direction one finds that the level of the peneplain drops from about 2500 metres at the Ruakwataro to some 2400 metres SW. of the Kaferongo (Berarara, Kagungu), and to 2050 metres on the next hill ridge (see fig. 16). The same is observed when proceeding from the Kamena (2390 m.) over the Kasooni ridge (2260 m.) and the Mukungu (2250 m.) towards the next ridge, the highest peak of which is 2139 metres. Finally the same dip is found farther to

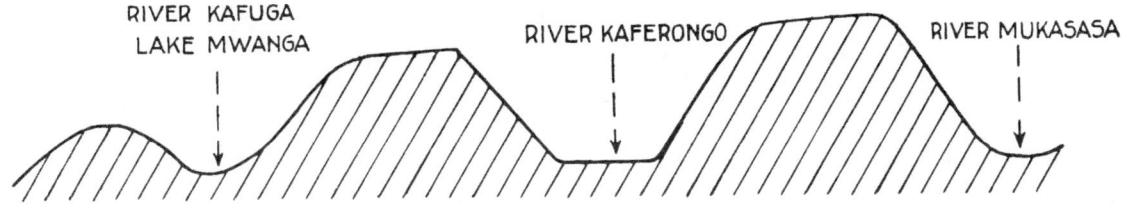

RIVER KAFUGA
LAKE MWANGA   RIVER KAFERONGO   RIVER MUKASASA

Fig. 16. Evidence of warping: Remnants of peneplain on ridges between rivers, decreasing in altitude to the south-west.

the NW, from 2413—2423 metres at Naiguru and Kasatora to 2150 m. at Heinoni and 2050 m. SW. of the Kafuga.

The peneplain, shown by the above figures to be fairly flat in a NW—SE direction, has undergone a folding in consequence of which it slopes down towards the Bufumbira fault-trough, thus to the SW. The consequent course of the Ruezaminda is due to the drowned valleys having overflowed to the SW. A large flood of water assisted by the considerable declivity tended to bring about a quick adjustment, and the Ruezaminda now flows through a broad and deep cañon (see photo 12).

In Kigezi there is a rain centre, the origin of which is probably due to the peculiar meteorological conditions prevailing between the Ruwenzori, the Virunga volcanoes and the humid Edward Rift. In this

Photo 12. Gorge of Ruezaminda facing west near confluence with Kaferongo.

district it rains almost every day, and as a natural consequence there is a wet, tropical forest, the Kayonza Forest, extending across the peneplain (see photo 13). In this forest one is struck by the fact that all the streams have their origin in swamp divides; pools of mud lying in narrow valleys which continue unbroken from the SE. to the SW.

In Kigezi all pre-miocene alluvial deposits will practically have disappeared. Only the upper Kanyamagogo, the Kirururuma and the Emise are likely still to contain some alluvial deposits. For the accumulation

Photo 13. View from Naiguru facing east, showing steep valleys cut in peneplain and covered with forest (Kayonza forest).

of cassiterite, therefore, the district is most unfavourable, from the geological as well as the morphological point of view.

Photo 14. View of Bunyoni near Muko. In the foreground lava, with outlet of drowned valley system to the left.

The Rubanda arena and the Luambogo valley are likewise drowned, since the backward erosion of the Ruezaminda has not yet advanced far enough.

Flows of lava, too, have affected the rivers of Kigezi. Lake Bunyoni is the drowned valley system of the Mukasasa (photo 14). A volcano was formed in the valley west of Muko and obstructed the passage. The level of Lake Bunyoni rose and its water found an outlet to the Luambogo. A later flow of lava threatened to shut off also that outlet, but the water of the lake continued to flow over the lava to the Luambogo.

Something similar to this also happened to Lake Mwanga.

At the time that these streams of lava — which are also found in the Rubanda arena — flowed out, the topography was exactly as it is now.

PART II

REGIONAL GEOLOGY.

CHAPTER 1

WISHIKATWA

The author made his first camp in Wishikatwa, near the village of Kagugu at the foot of the Ruitonbero, a spot which proved to be an ideal place for a study of the Lower Division of the Karagwe-Ankolian system. The camp was a short distance west of a small hill top (No 18), where the outcrop was discovered of a metamorphic quartzite (cf. photo 52, part IVB, chapter 3) with mica and tourmaline on the bedding planes.

This quartzite, gently dipping towards the west ($\pm$ 18°—25°) is resting upon gneissose biotite granite, which away from the sediments merges into massive, coarse-grained, leucocratic muscovite granite. This latter variety is commonly called "pegmatite granite". The above mentioned quartzite is 5—10 metres thick and is overlain by tourmaliniferous, quartzose mica-schists. The degree of metamorphism of these schists is not at all uniform; with a magnifying glass small lenses of white or bluish sericite are always to be found in hand specimens and the sericite sometimes constitutes such an important part of the rock that the latter deserves to be called tourmaliniferous, micaceous and partly silicified sericite phyllite. Exploration right down to the foot of the Ruitonbero (No 17) is rendered impossible by detritus. The Ruitonbero is formed of dark reddish brown, well stratified and thickly banked rocks crowned by a thick, metamorphic quartzite.

The reddish brown strata are sandy phyllites, consisting of quartz, grey silky sericite, darkish brown, fine-scaly mica, hematite and very numerous garnets, the latter generally well-shaped though rather strongly weathered; in diameter they vary from 5 mm to 2 cm. These phyllites have been called garnet phyllites and are dipping towards the west, the angle of dip being approximately 25°—28°. Granite again is found west of the overlying quartzite, which latter quartzite is hereinafter called the "Ihunga quartzite" — Q_2, according to Combe.

In the cross-faults intersecting this series south of the Ruitonbero, pipe-like masses of a chocolate coloured rock consisting of tremolite and chalcedone are found.

As the sedimentary strata are devoid of fossils, the only way to map the structure of this landscape is to follow a certain horizon in the direction of the strike, since of course tectonics and stratigraphy are inseparable.

The series of the Ruitonbero was therefore traced to the north, but it soon dies out in granite, whilst the Ihunga quartzite extends farther to the north than the other horizons. However, this quartzite also disappears and its former extension is only evident from a larger quartz content of the granite. The lowermost quartzite was followed along the direction of the strike first towards the south, then eastward and, though a transverse fault may exist at the foot of the Kavusanammi-East, it can easily be traced as far as the Omutarraz (hill No 26). From there again it can be traced southward till fairly close to the Katuba. This quartzite was invariably found to be resting upon granite, and for this reason the present author has called it the "boundary quartzite", which proved to be a very good name (Q$_1$ of Combe).

Near the Omutarraz (hill No 26, cf. maps 4 and 5), the boundary quartzite was overlain by bluish-grey, sometimes variegated, silky sericite phyllites, which close to the granite, thus immediately above the boundary quartzite, in many places change into tourmaliniferous mica-schists, which latter also occur farther from the granite as lenticular enclosures in the unchanged phyllites.

The arenaceous strata of the next overlying narrow horizon consist solely of quartz and mica and in some places contain many garnets, whilst they frequently show a typical zig-zag foliation. This horizon is not everywhere developed, and, of course, even where it is present it is not always exposed (as for instance at the foot of the Ruitonbero). It might, therefore, seem as if its large quartz content were due to local, intense additive metamorphism, but neither the upper part of the sericite phyllites nor the overlying garnet phyllites show signs of noticeable contact metamorphism. The fact that in some places a garnetiferous quartzite is found intercalated between the sericite phyllites and the garnet phyllites, explains the occurrence of quartz-rich, micaceous strata between these two horizons in other places. Thus, near the Ruitonbero as well as near the Omutarraz, a quite similar succession of strata is found, in both cases dipping away from the granite; this succession is formed by the boundary quartzite, the "upper sericite phyllites", a narrow arenaceous horizon and the garnet phyllites, which succession on the Ruitonbero is crowned by the Ihunga quartzite. Whilst the extension of the boundary quartzite could be traced easily, in spite of a transverse fault, from the Ruitonbero up to the Omutarraz, this is no longer possible when following one of the higher horizons.

The garnet phyllites for instance extend from the Ruitonbero as far south as the Kavusanammi-East, but there they abut at right angles against the belt of upper sericite phyllites running from the Omutarraz towards the west.

The strike of these sericite phyllites is invariably E-W. Their angle of dip is very small on the Omutarraz, but increases gradually towards the west, till on the northern spurs of the Kavusanammi-West the strata are locally over-tilted. Starting on the Omutarraz and following the garnet phyllites one consequently does not arrive at the Ruitonbero, but reaches the southern spurs of the Kavusanammi-West. A big disturbance undoubtedly exists across the top of the Kavusanammi-East, apparently increasing in importance towards the west.

Along this disturbance one may notice abnormal contacts between different horizons, and different directions of strike are found on both sides of the fault-plane. Overtilting of strata, which is relatively rare in S.W. Uganda, occurs to the south of this remarkable fault. Moreover the sericite phyllites that extend westward from the Kavusanammi-East via the Kavusanammi-West up to the Ihunga contain coarse crystals of cyanite and well-shaped crystals of staurolite, both of which minerals are quite unusual for the sericite phyllites. Their formation may be partly due to the presence of granite to the north of this zone of sericite phyllites, but the main cause of their genesis must be sought in the tectonics. The explanation of all these peculiarities will be referred to in the next part. However, it may here be emphasized that cyanite is a very rare mineral, whilst staurolite has never been found in sericite phyllites except in the above mentioned zone [1]). Following the western water-divide of the Katuba river from the Kavusanammi-West to the south one finds the garnet phyllites overlain by a strongly metamorphic quartzite, undoubtedly the Ihunga quartzite, in an almost vertical position (hill No 31). In some places there is a little mica on the bedding planes, but the characteristic feature is the replacement of thin alternating layers by tourmaline. Coarse crystals of tourmaline have also been found in the underlying, uppermost layers of the garnet phyllites. Proceeding towards the Chamiombu (via hill No 33) the garnet phyllites reappear, here with an easterly dip, the strike having gradually changed from E-W to NNW-SSE.

Near the top of the Chamiombu (hill No 34) the grey sericite phyllites are again exposed, underlying the garnet phyllites, and at the top of the steep SW. slope of the Chamiombu a quartzite is found.

Thus when going over the Kavusanammi-West towards the Chamiombu a syncline is crossed, which near the axis shows evidences of contact metamorphism.

Going down from the Chamiombu towards the Lugalama arena one finds that here the boundary quartzite is not immediately resting upon granite, but that granite and quartzite are separated by an intercalation of sericite phyllites

---

[1]) It is a common contact-metamorphic mineral in clayey-arenaceous phyllites, when in close vicinity to the granite. It is found for instance in the garnet phyllites on the Ruitonbero, where these abut against the granite, and it is also found in Kavungo in arenaceous, phyllitic shales of the Middle Division in the same circumstances.

which are exactly similar to those stratigraphically above the boundary quartzite and which are called the "lower sericite phyllites".

Proceeding eastward from the Chamiombu over the ridge (cf. section 1)
again a syncline is traversed. The garnet phyllites become gradually steeper
but retain an easterly dip. A few hard varieties (near hill 33) were found to
contain, in addition to garnets and quartz, staurolite and cyanite, whilst
remarkably enough mica was practically absent [1]). Consequently again here
indications of stronger metamorphism are found in the axis of a syncline.
Towards the east grey, silky sericite phyllites are again encountered, in the
midst of which a quartzite is found, the dip of which is about 80° in an easterly
direction. This quartzite runs with a NNE-SSW strike from the Nyihanga
towards the north till it abruptly dies out on the northern slope of hill No 36;
it is undoubtedly the boundary quartzite. Descending from the outcrop of
this quartzite towards the arena of Rushenyi, it may be noticed that the
direction of dip of the overtilted lower sericite phyllites is reversed twice,
whilst finally the angle of dip decreases. At the foot of the arena
wall a slight elevation of the surface is caused by the outcrop of a quartzite
gently dipping towards the east. It is overlain by sericite phyllites, which soon
make place for granite. Evidently an anticline is crossed, steeply overtilted
towards the west, as is also the syncline of the Chamiombu. This anticline, for
easy reference, will be called the Shonobutondo anticline (cf. also section 2 and
map 5). When proceeding eastward from hill No 36, following the ridge towards
the Shonobutondo (hill No 39), an abrupt change in the direction of strike is
noticed, the direction changing from NNE-SSW into WSW-ENE (hill a, map
5). At the same time a number of quartzite fragments are encountered immediately behind the boundary quartzite; these fragments have been laid down
in the manner of roofing tiles and mostly have a steep southerly dip. A common feature is that every fragment lies south of its easterly neighbour, and
this phenomenon can only be explained by assuming that the compressional
force did not always act in the same direction. These fragments are certainly
parts of the boundary quartzite, as they are lying in sericite phyllites. Moreover,
the fifth fragment to the east, which has a northerly dip, is overlain by sericite
phyllites underlying garnet phyllites: consequently these fragments belong to
the southern flank of the Katuba syncline (or the northern flank of the
Shonobutondo anticline). In the latter anticline granite is exposed here and
there, whilst a big disturbance is found in the southern flank between hills 38
and 42. Section 2 shows the exposures met with when crossing the Katuba
valley from hill 37 towards hill 27.

This very same succession of sericite phyllites — boundary quartzite—sericite
phyllites — garnet phyllites is found when proceeding from the small hilltop NE
of the Omutarraz (hill No 26) towards the Nabusov (No 23), and once more in the
inverse order when passing over from the latter hill to hill No 22. Thus another
syncline is crossed, which fact is also evident from the directions of the dips.

1) Cf. photo 24, part III.

CHAPTER 2

DEVELOPMENT OF THE LOWER DIVISION [1]) OUTSIDE WISHIKATWA

On map 4 it may be seen that the boundary quartzite of the Chamiombu encircles the whole of the Lugalama arena, sloping away from it on every side. Section 3 shows the succession found when crossing from this arena to the Ihunga region, and it is seen that it is the same succession as previously described. The only difference is that the arenaceous mica schists between the upper sericite phyllites and the garnet phyllites are well developed, especially on the north flank, whilst thin layers of sandstone are clearly seen in this horizon. Whereas on the south flank many garnets still occur in the garnet phyllites, in the corresponding horizon on the north flank they are very scarce, the reddish-brown phyllites on that side being more arenaceous.

The north flank of this syncline was also followed from the Kavusanammi-West (No 30). After crossing the Kachwamba one first finds on the Mushash (No 11) a remnant of the boundary quartzite, a strongly metamorphic fragment of which was seen on the Kavusanammi-East (No 29). This north flank is everywhere overtilted, also on the Mushash, the Kakanenne (No 10) and the Ruacherenze (No 8), whilst cyanite is particularly abundant in the sericite phyllites. The southern flank may easily be traced from the Chamiombu (No 34) via the Nerionza (No 32) and the Mushash up to the Chamter (No 70), to the west of which latter hill the normal succession is found with boundary quartzite and with garnetiferous garnet phyllites. To the west of these phyllites lies a disturbed triangular area which, to make matters still worse, is for the greater part covered by valley floor detritus or alluvia from the Nya Ruambu. The north flank of the first Rukiga syncline forms the southern side of the Nya Lusanje valley and at the same time the southern wall of the Lugalama arena.

The question in what way the monocline of the Chamter is related to the north flank of the first Rukiga syncline has not been solved. Ascending the Rukiga hills from the valley to the south of the Ihunga, one finds the normal succession of the Lower Division dipping towards the S.W.: sericite phyllites, garnet quartzite, garnet phyllites and Ihunga quartzite. The same succession is observed when ascending from the arena of Lugalama. To the west of the garnet phyllites of the Chamter, however, one finds highly disturbed sericite phyllites. Consequently there must be a considerable fault to the west of the Chamter, probably dying out in the steep escarpment formed by the Rukiga mountains.

Presumably no fault exists between this escarpment and the Ihunga, but between the Ihunga and the Chamter there must be one.

---

1) Thickness about 1500 M.

As seen from section 4, the Ihunga forms an intermediary syncline (inversion of the topography!) between the granite and the first Ruigka syncline, the axis of which runs approximately W—E. As the north flank of the Ihunga is the immediate continuation of the north flank of the syncline encircling the Lugalama arena, the fault between the Ihunga and the Chamter must die out somewhere in the sericite phyllites of this north flank.

The few outcrops found between the foot of the Ihunga and the Nya Ruambu river give evidences of the presence of disturbances.

Owing to the great height of the Ihunga a thick quartzite is found on the top, which is not usual in Wishikatwa or Rushenyi. This is underlain by the garnet phyllites, which, here, are entirely devoid of garnets. These phyllites are red and arenaceous to clayey, with small conglomeratic intercalations at the top; in appearance they are exactly similar to those occurring SE. of the Ruacherenze (on hill No 9). Underlying these phyllites is a fairly thick quartzite layer with a very large number of cavities of about the same size as the garnets occurring in the garnet phyllites elsewhere. Below this layer are greyish-blue sericite phyllites lying immediately upon the gneissoid biotite granite and forming the direct continuation of the corresponding belt extending from the Mushash (No 11) over the Kakanenne to the Ruacherenze (No 8).

The boundary quartzite is undoubtedly absent in the Ihunga, possibly due to the intrusion. The uppermost quartzite is undoubtedly the equivalent of the quartzite on the Ruitonbero, and the writer has in fact named this the Ihunga quartzite (designated by Combe as Q 2). It is remarkable that the dips, which on the top are not too steep (up to 50 degrees), going down the eastern slope become steeper and steeper. The quartzite characterized by cavities is without doubt the equivalent of the arenaceous horizon shown in section 3 and described in the previous chapter; also on the valley side between Kabuye (No 77) and Nya Lusanje (No 6) it is very clearly seen and is to be taken to belong to the Rukiga facies of the system.

Neither this "garnet quartzite" nor the accompanying arenaceous horizon extends to the east, and their absence between the eastern foot of the Ihunga and the eastern banks of the Nya Ruamba must be due to the presence of two faults, the western one of which forms the eastern boundary of the Ihunga syncline.

In the west the Ihunga is likewise bounded by a fault, the land to the west of which is formed by the sericite phyllites. The Ihunga, therefore, is a minor fold bounded by transverse faults and shows higher horizons than those of the typical synclines wedged in between granite bodies. This also shows that the same synclines on a higher level dip much more gently than is the case on a tectonically deeper level (cf. sections 2 and 3 with section 4).

The boundary quartzite is likewise absent in the southern flank of a syncline situated to the north of the Ihunga and separated from this latter mountain by granite. It may, therefore, be doubted whether the boundary quartzite has ever been present in these two synclines, the more northerly of which is called the Ruamheji syncline (hill No 4).

This syncline is shown in section 5 and does not call for much comment, except that the north flank, on which fragments of the boundary quartzite abut against the granite, is in some places overtilted. The garnet quartzite (designated by Combe as Q 2a) is not found in this syncline, whilst the garnet phyllites show the Ihunga facies.

A remarkable feature is the occurrence of a thin intercalation of strongly weathered feldspar-porphyry rich in quartz at the base of the garnet phyllites in the south flank, although otherwise effusive rocks are entirely absent in the system. Possibly this is a dyke. The feldspars are completely altered into sericite and the ground mass consists of fine-grained quartz.

The fact that on the Mushash (No 11) and the Mtunda (No 3, cf. map 4) small remnants of the boundary quartzite are found to be intercalated between the granite and the sediments, supports the view that this quartzite has disappeared in consequence of the intrusion. On the other hand the belt of sericite phyllites on both sides of the Ihunga, as well as in the south flank of the Ruamheji syncline, is very thick, so that one is inclined to assume that these phyllites comprise both the lower and the upper sericite phyllites.

Possibly both explanations are right and the assimilation of the boundary quartzite by granite near the Mushash and north of the Ruamheji may have been favoured by the fact that it is thinning out towards the Ihunga and the Rukiga mountains.

This boundary quartzite seems to increase in thickness towards the east. It is certainly present around the Chitwe arena (cf. map 3) and the author has found traces of it near Rusinga, Nanyankoko and Kikagati. It probably encircles also the Ntungamo granite, as it does the Rushenyi arena.

It possibly covers large areas in Bukanga, judging from Miss Ledeboer's map. The author was only once in a position to visit this area and he was then struck by the remarkable likeness of the Bukanga phyllites to the non-metamorphic phyllites of the Lower Karagwe. As, moreover, the almost horizontal quartzite of Bukanga appears to be stratigraphically lower than the quartzite of Nsongezi, which is undoubtedly the Ihunga quartzite, we may safely correlate the Bukanga quartzite with the boundary quartzite (cf. section on map 3).

From the account given by Salée it is to be concluded that this quartzite also occurs near Gatsibu, whilst according to sections surveyed by Miss Ledeboer it is likely to be present in southern Maziba quite near the contact with the granite.

Whilst the boundary quartzite thins out to the west the Ihunga quartzite (Combe's Q 2), which forms the upper boundary of the Lower Division, increases in thickness westward. The latter forms the top and western slope of the Ruitonbero, is met with in the core of the Katuba syncline (see map 5) in a tourmalinized state, and crowns the Ihunga. Whereas according to Combe the Ihunga quartzite near Mwirisandu is no more than some 30 metres thick, its thickness in the Rukiga hills south of the Ihunga is about 150 metres. This quartzite gives a good indication of the general trend of the folding of the Karagwe-Ankolian system. Miss Ledeboer discovered it north of the Chitwe arena, where it overlies Q 1 and takes a sharp bend to the south, forming the northern arena wall of the Ruberogoto arena and crowning the valley sides of the Kivimbiri valley, before crossing the Kagera near Kikagati. Both Combe's and Salée's maps agree in indicating this Ihunga quartzite as encircling the Ibanda arena, and probably also that of Kyerwa.

Along the Kagera the Ihunga quartzite returns in the east flank of the Karagwe syncline near Nsongezi, and the road leading to Mbarara has to make a wide detour to pass over this quartzite via a windgap. In that part the author had several opportunities to observe the sediments of Bukanga, and in every case he was again struck by the great similarity with those of Wishikatwa, in respect of the general habitus.

The Ihunga quartzite is clearly seen in the vicinity of Nya Lusanje, crowning the steep mountain slope extending to south of the Chamter and forming the high southern arena wall of the Lugalama granite. Combe has mapped the scattered fragments of this quartzite near Lutobo and, judging from the sections given, Miss Ledeboer encountered it near the Muvumba valley in the Gombolola Maziba (Kigezi).

The author believes to have recognised it east of the Rubanda arena and also on either side of the axis of the Rubanda anticline on the north side of the Kayonza Forest; between these places, near the Nyakashunzu (close to the river Tsyaga, near Kasyoji village), he again found the lower Karagwe with sericite phyllites, in places tourmalinized, as also pure quartzites and garnet phyllites.

The occurrence of the garnet quartzite is very erratic. It is present in the Ihunga and to the south of it, but not in the Ruamheji syncline. It is again found in the Buramma syncline, which separates the Rushenyi arena from the Ndorwa arena (cf. section 6). In the latter syncline there is a combination of the Rukiga and the Wishikatwa facies; the boundary quartzite is well developed and the garnet phyllites contain very many garnets, but on the other hand one finds the garnet quartzite well developed as one or two layers of garnetiferous quartzite in a narrow belt of slightly micaceous sandstones.

On the whole, however, it seems as if the Lower Division is much thinner here than in Wishikatwa and near the Ihunga.

CHAPTER 3

MIDDLE [1]) AND UPPER DIVISIONS OF THE KARAGWE-ANKOLIAN SYSTEM
(SEE MAP 6)

In this chapter particularly the geology of the Kigezi district will be discussed, as this district is chiefly built up by the higher horizons of the Karagwe-Ankolian system. There are still large gaps in our knowledge of this district and the author's map 6 should be regarded as a tentative representation of the structure of Kigezi. This district presents great difficulties to geological exploration, partly because of its trying climate and partly on account of the dense vegetation. S. J. Vermaes has explored the triangular area between

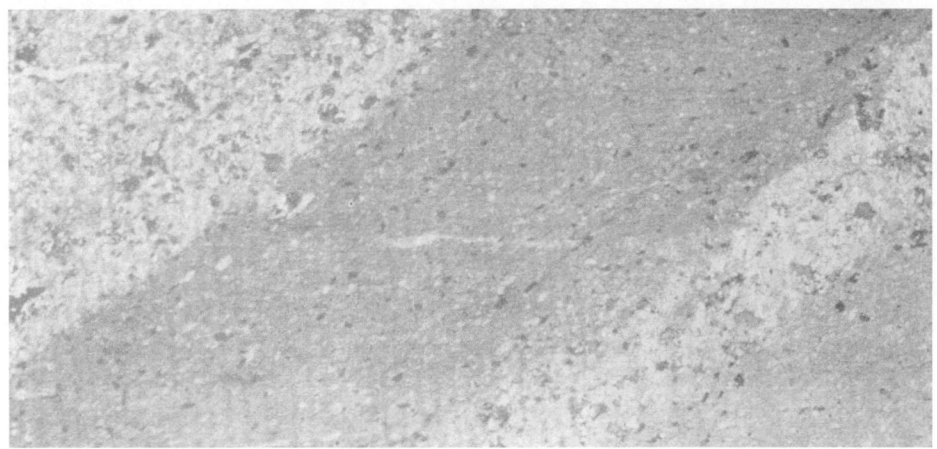

Photo 15. No 107 Kigezi (ordinary light, magn. 14x). Phyllitic shale of Middle Division. The foliation is parallel to the longer edge of the photo.

Mafuga, Kumba and Nya Lusanje. Combe has published his survey of a small part of Kigezi along the Kanyamagogo, adjoining Ankole. The author has roughly mapped the area bounded by a line through Muko, Nya Kalembe and Lubuguli up to near the Congo frontier, thence eastward to north of the Kayonza forest and the Rubanda arena as far as Kumba.

Roughly speaking, the Middle Division is built up by more or less arenaceous phyllitic shales, mostly of a bluish or reddish colour. They often show a torrential bedding and their original bedding planes often make a wide angle with the foliation (cf. photo 15).

The sediments become more arenaceous towards the top and are overlain by a thick horizon consisting of quartzitic sandstones (cf. photo 16) and yellow micaceous sandstones or sandy micaceous shales. This is the most important key horizon in Kigezi, and it often contains conglomeratic intercalations and also real itabarites. This "sandstone horizon" is overlain by mostly clayey, phyllitic shales of the Upper Division, which by weathering yield a very fat

---

1) Thickness estimated by Combe at about 4000 metres in E. Kigezi.
Stheeman, Geology.

clay; towards the top these become more arenaceous. A single quartzitic sandstone is found to occur in this division above the sandstone horizon.

These clay-shales sometimes contain relatively large mica scales, which certainly are not of sedimentary origin (cf. photo 17), as they are almost perpendicular to the stratification. These mica scales are much bent and contorted and the shales must have been strongly compressed during the folding.

At some places itabarites have been found in the more or less arenaceous shales

Photo 16. Quartzitic sandstone, Kigezi (crossed nicols—magn. 14x), from sandstone horizon. The indented outline of the coarse grains has originated from the incorporation of the siliceous part of the cement. The aluminous part of the cement consists of sericite.

constituting the uppermost part of this division so far as known.

The sandy phyllitic shales of the Middle Division have been seen by the author along the road from Lutobo to Kabale or from Lutobo via Soko to Nya Lusanje, between the Ihunga quartzite and the sandstone horizon. From east to west this latter horizon shows itself repeatedly and in this country with

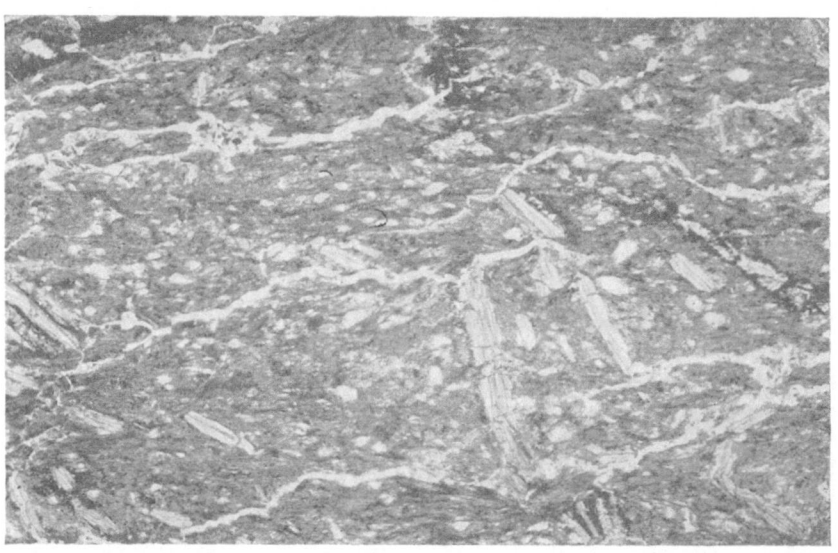

Photo 17. No. 109b, Kigezi (ordinary light, magn. 31x). Path Muko — Rubanda. Micaceous phyllite of Upper Division. The slide has been cut almost parallel to the stratification. The mica scales, which are bent and broken, are lying approximately perpendicular to the stratification.

its rather dense vegetation it gives the best outcrops.

To the west of the Kanyamagogo valley this horizon appears again, and is seen for the third time near Kabale [1]). Proceeding from Nya Lusanje to Kumba one sees the Ihunga quartzite overlain by the thick belt of phyllitic or sandy micaceous shales, and just west of the Kanyamagogo valley this in turn is overlain by the westerly-dipping sandstone horizon. Vermaes' map shows that between Kumba, where the sandstone horizon is found to have an easterly dip, and the Kanyamagogo valley, the mountain land is formed only by phyllitic shales. Here, therefore, only one syncline is found, whilst between Kabale and Lutobo there are certainly two. The intermediate anticline must, naturally, die out somewhere. This has been schematically indicated by the author on map 3.

As said before, two features characterize this sandstone horizon. In the first place it is often found to contain conglomerates. These have been found both in the Kanyamagogo valley and at Kabale and Kumba, whilst Vermaes has observed them near Mafuga. The pebbles from these conglomeratic inter-calations do not differ in appearance from the rocks of the Karagwe-Ankolian sediments, though it is certainly to be assumed that at the time of their depo-sition these pebbles consisted of hardened sediments. Quartz pebbles, however, are by far the most predominant. Unfortunately the specimens collected by the author were lost on his journey through the Kayonza forest, so that no petro-graphical examination could be made. These specimens were taken on the west side of the road from Kumba to Kabale, about half-way, where the Rubanda path turns off from the main road and the direction of the Kiruruma valley changes from NW. to N. There the sandstone horizon forms both sides of the valley and the strike is about 150 degrees, the slope on the western valley side being about 50 degrees to the east. Proceeding from that place up to the road to Kumba one sees the dip change to a steep westerly one until near Kumba, where the sand-stone dips steeply to the east again. It is probable that here the sandstone horizon, which is very broad at that part, forms the axis of an outcropping syncline. Beyond Kumba the valley of the Ishasha breaks through the sandstone horizon in a westerly direction, this horizon continuing NNW, in which direction it has been followed by Vermaes to beyond Mafuga.

The second feature of the sandstone horizon, viz. the large content of hematite, was likewise found to be constant, whilst in some places real ita-barites, consisting of hematite and magnetite, occur in the upper portion of

---

1) It has been contended that the Karagwe-Ankolian system in Kigezi has been developed into a more arenaceous facies. This is undoubtedly erroneous, the mistake being due to the repeated occurrence of the sandstone horizon. Maybe the sandstone horizon in Kigezi is much thicker and better developed than in Karagwe, but the contrast between this horizon and the underlying and overlying predominantly argilla-ceous shales continues to be just as pronounced. Combe does not indicate this horizon as a whole; he shows the individual sandstones and quartzites as Q3, Q4, and Q5, both on the hill ridge separating the Ndorwa arena from the Ibanda arena and in the valley of the Kanyamagogo. The quartzite Q6 is taken by the dresent author to belong to the Upper Division.

this horizon. This, too, has been noticed by Vermaes near Mafuga and east of Buchundula.

The west flank of the Kiruruma-Emise valley syncline is cut by the motor road from Kabale to Chabahinga (the ferry place over lake Bunyoni), along which road the steeply east-dipping sandstone horizon is clearly seen. Close to the lake shore dark-red, thinly banked, phyllitic shales of the Middle Division are seen to dip eastward. There we are in the Bunyoni-Rubanda anticline, in which granite is again exposed.

On the opposite side of the lake the sandstone horizon reappears in the western flank of the anticline, in which it continues in a northerly direction up to the sharp westerly bend of the lake. Here it crosses the lake and takes a more north-westerly strike, thus forming the southern wall of the Rubanda arena, where it is clearly exposed along the Muko-Rubanda road.

The Muko rest-camp on the western point of the lake lies on a spur of the high hills to the south, which spur consists of thinly banked and well stratified phyllitic shales of the Upper Division alternating in colour from dark reddish brown to bluish grey, yielding in the weathered state a very fatty impermeable clay.

Going down towards the valley one first crosses a lava flow consisting of a vesicular basalt which flowed out at a time when the topography was already exactly as it is now; old sediments do not occur again until beyond the present outlet of the lake in the valley of the Luambogo against the eastern valley slope. These sediments are more or less sandy phyllitic shales with a strike of about 115 degrees, whilst farther eastward along the lake the strike is already 140 degrees; there is, therefore, a marked bend to the WNW.

Where the path through the swampy valley of the Luambogo turns off to the NE, quartzite makes its first appearance. The strike of the numerous successive quartzites and sandstones varies between 115 and 135 degrees, whilst the dip is invariably SW. Past the watershed between the stream flowing to the Luambogo and the Kamonyo (farther on called the Wlamiora) the quartzites disappear and are succeeded by yellow micaceous sandstones. Emerging rather suddenly from the hills one then enters the swampy Rubanda arena; ridges of granite separate the drowned valleys and hummocks of bare granite are visible at a great distance. The NE. slope of the arena wall is already formed by granite, whilst a peculiar dynamometamorphic contact rock is found between the sediments and the granite. The last sandstone merges into this contact rock. The dip is S. and the strike here is 110 degrees, thus fairly parallel to the boundary of the granite.

In the contact rock and close to the granite is a quartz vein containing baryte, pyrite, pennine and mica, and a few small prisms of tourmaline, which is very pleochroic from almost white to greenish black. The path then enters the arena and leads almost directly on to a basalt very similar to that of Muko; this basalt has flowed out into the now drowned river valleys, and, where the lava stands out above the present level of the swamp, valleys can always be forded dry.

Farther westward the surface of the lava [1]) disappears below the level of the swamps. Continuing one's way from Rubanda to Kumba, sediments are again encountered just before reaching the wind gap between the Kaban-damma and the Nya Melengere. These sediments consist of phyllitic shales, but just beyond the wind gap two quartzites are found in quick succession, extending over the Katoma to the Wramuyonyi, now standing vertical, then again dipping westward or eastward. The path then crosses a thick belt of phyllitic shales, sometimes alternating with sandy phyllitic shales. Approaching close to Kumba the sediments become very clayey and strongly coloured, and in these phyllites, as also along the Bunyoni lake, one finds a number of patches devoid of vegetation, which patches are favourite licking spots for animals.

The strike of the sandstone horizon, which is then found just before Kumba with an easterly slope, is rather N-S. Thus it is seen that the syncline deviates more or less to avoid the Rubanda granite boss. Still clearer, however, is this to be seen in the phyllites of the Middle Division, which at the confluence of the Nechanga and Nya Magama have a strike of about 145 degrees, whilst between the wind gap and Kumba the strike is practically N-S. Farther on the strata turn again more westerly, the sandstone horizon near Mafuga running approximately NW, whilst the phyllitic shales of the Middle Division make a still sharper bend. On climbing the Wramuyonyi on the north side of the Rubanda arena one sees that the strike at the foot of the mountain is about 115 degrees, whilst that at the top is round about 140. On the steep and bare mountain slope small minor folds are clearly visible, some overtilted to the SW. Detail folding is possibly the explanation of the varying thicknesses found in the horizons in Kigezi. On the way from the Wramuyonyi to the Katoma there very soon reappears the quartzite that was noticed on the Rubanda-Kumba road just after leaving the wind gap. If, as the writer assumes, this quartzite is to be taken for the Ihunga quartzite, then the underlying phyllites and sandy phyllitic shales would belong to the Lower Division and the strata on the northern slope of the Wramuyonyi would be the equivalent of the garnet phyllites; however, garnets are lacking, as was also found to be the case on the Ihunga, but judging from the general habitus this assumption is by no means impossible. The Ihunga quartzite was followed by the author via the Katoma and the Kabandamma to the Nya Melengere, and it is quite possible that the water divide a little SE. of the Molusetsye is formed by this quartzite. The vegetation, however, considerably impedes geological exploration and it could not be established whether or not the strata at the Mwanya Mehanga on the SE.

---

1) This lava once emanated from the now extinct Tyekombe volcano; the eruptions must have been principally on the eastern side, since the level of the lava, which reaches to the eastern arena wall, slopes down from that side along the north and also the south side of the volcano, the ridges between the rivers on he west side being no longer covered.

The salt market of Rubanda is situated on one of these lava-filled valleys.

point of the Rubanda arena sweep around the granite; the writer believes that they do not do so. They certainly do not do so at the northwestern corner of the Rubanda granite. This important question was borne in mind when exploring the western spur of the Kashuli, and in fact immediately west of the Kitauölira, which is formed of granite, a strike of 150 degrees and a dip of 45 degrees west are found. Also on the western flank of the Kashuli there is no trace of the strata sweeping around, though they do indeed seem to have been disturbed, which, as a matter of fact, is only to be expected here, as also in the SE. corner of the arena.

A little way back from Rubanda to Muko the lowest sandstone of the sandstone horizon was observed overlying the contact rocks of the Rubanda boss, and in these contact rocks, as already remarked, quartz lenses occur which contain i.a. baryte and pyrite. The author followed up these pyritic lenses and found them to occur south of Luamebale, on the Habubali, on the island of Muga and also on the spur of the Kashuli SE. of Kasyanda village. Thus, this row of lenses very accurately sweeps around the granite. The sandstone horizon runs via the Kagoy over the Lwankuba to the Masiazu, which latter is formed by a sharply protruding quartzite (masiazu = sharp); there the strike is 127 degrees and the dip 77° SW. Whereas the Habubali is for the main part formed by granite, the granite boundary sweeps around in a northerly direction, cuts the island NW. of the Habubali [1]) and then passes east of the island of Muga and the three islands farther eastward. These islands, therefore, are built up from sedimentary rocks, and the strike is 150 degrees with a westerly dip of 50—60 degrees. On the island of Muga the strike is 175 degrees, so that also here the sandstone horizon more or less follows the boundary of the Rubanda granite. Then the sandstone horizon, however, continues west of Endegwe village, turning away from the granite in a NW. direction.

This horizon is underlain by the clayey shales of the Middle Division, under which the Ihunga quartzite would be expected, and indeed a quartzite is found east of Endegwe village in the valley of the Tsyekurwa, with a strike of 145 degrees and a westerly dip. The valley of the Tsyekurwa was then followed as far as the fork, where the western side was climbed up to the crest. At first phyllitic shales were found, then sandy shales and on top of those micaceous sandstones. Finally a quartzite was reached, presumably the continuation of that found at the entrance to the Tsyekurwa valley. The crest was covered with a fairly thick layer of earth and leaves, which made it difficult to continue the exploration. Descending along a branch valley of the Ishasha, phyllitic rocks were found which produced a very heavy clay and showed strikes of 150 to 165 degrees with a steep westerly dip. The river was then crossed and an exploration made around the water divide of the next stream. There for the first time quartz lenses were again encountered, and also sandy phyllites

---

[1]) All these names are connected with rocks or stones, called in the Lukiga language "ama*bari*", the granite here weathering to large boulders, the so-called "Wollsäcke".

and sandstone, which had a strike of 155 degrees and a steep easterly dip.

The last mentioned sandstone was again found in the valley of the Bitanwa, with a steep easterly dip (70 degrees), while the strike was about 160 degrees. The hill ridge was then followed and the descent made on the west side with the intention of reaching the Mbwa valley, but by mistake a more northerly spur was taken and the author had to camp in the Tsyaga valley near the village of Kasyoji. Here sericite phyllites with tourmaline were found, and also a few garnets, certain evidence of the presence of the Lower Karagwe, so that in all probability the quartzite seen in the Bitanwa valley was the continuation of that of Kabandamma-Katoma-Wramuyonyi, thus the equivalent of the Ihunga quartzite in the eastern flank of the Rubanda-Bunyoni syncline. The strikes at Kashoji were about 160 degrees.

Next the Nya Kashunzu was climbed and the western continuation followed. Here the rocks were difficult to trace owing to high ferns and bush, but several quartzites were observed to run at a small angle over this ridge.

Quartz veins containing arsenopyrite were also encountered.

Opposite the Dwanzu the descent was begun towards the Mbwa valley. The slope was again found to be constituted mainly of more or less sandy phyllites, whilst on the Dwanzu red sandy or phyllitic shales were found, the strike on both sides being 140 degrees with a steep westerly dip. Dense forest was then met, but on the ridges many sandstone fragments were found; presumably the sandstone horizon is situated there. For the rest of the journey the slopes all consisted of rocks yielding heavy clay, making the way a precarious one.

The passage from the Dwanzu to the Kasatora or the Naiguru is most monotonous[1]), until on the flanks of these mountains a series of quartzitic sandstones are found alternating with micaceous sandstones, sandy shales and itabarites. There the strike is 140 degrees and the dip 45 East. On the other side of the Lubuguli valley the same sandstone horizon with itabarites is still present.

Continuing northward around the Kara towards the Ruezaminda valley, one finds only the sandstone horizon, which again is encountered after crossing the last-mentioned river at the confluence with the Kaferongo and climbing the Mukungu. Also the Heinoni is formed of these sandstones and very clearly shows an easterly dip. Their great thickness here is accounted for by a minor fold, the axis of which just passes through the Kara; the strike on the Mukungu is 150 degrees and the dip about 45 NE.

The Kasooni, the Nya Kalembe and the following hills called Berara and Kagungu rise high above the valley of Lake Mwanga, owing to their being formed by the sandstone horizon. Just before reaching Nya Kalembe the strike turns in a direction more N-S. On this barren ridge it is clearly observed that this complex is formed of very many quartzitic sandstones and micaceous

---

1) The slopes are exceptionally steep; about 30—35 degrees. On these slopes one cannot but notice sandy rocks where these do occur, for there is a very sharp contrast between the slippery slopes, formed by clayey phyllitic shales, and those where sandier rocks outcrop, yielding a much more permeable detritus.

sandstones, and it is the writer's belief that it will no longer be possible to follow and recognize each individual layer. On this hill ridge there no longer occur the itabarites from which on the Naiguru and the Kara the natives obtain the iron for their spears.

From Nya Kalembe to Muko the path leads again through the Kaferongo valley, on the farther side of which one finds first phyllitic shales, then a single sandstone; this may correspond to Combe's Q6, which the author takes to occur in the Upper Division. Again south of the Luankuba and the Masiazu, on the ridge of the hills Ndolero, Kamioro and Bugomorre, phyllitic shales have been found with a single quartzitic sandtone, which continues on the one side to the Kara (spur of the Nya Kashetye = Mr Bamboo) via the Luembogo and thence to the Ihunga (SE. point of the Kayonza forest), and on the other side in the hills behind the Muko rest-camp and those south of Lake Bunyoni, where the strike is about 140 degrees and the dip SW.

Between the valley of the Kaferongo and Lake Bunyoni, clayey phyllitic shales of the Upper Division occur, mostly red in colour, but sometimes grey, with coarse mica folia. There is a slight overtilting to the SW. in the middle of the syncline on the Butali, and near the Munzugu there are sandy shales with a strike of 150 degrees, whilst also here itabarites are to be found. In the argillaceous phyllites of the Upper Division there are hardly any licking places.

CHAPTER 4

A TECHNICAL DISCUSSION OF THE MAPPING OF ARENAS

A study of sub-arid erosion is not only important for the tracing and evaluating of primary stanniferous deposits, but it is also of great practical value for the mapping of these regions. On comparing Prof. Salée's map with that of the author (map 3) and observing how the Ndorwa arena has been mapped, it will be seen that in the writer's opinion that arena is formed entirely of granite, whereas Salée indicates only some marginal outcrops of granite. Combe, too, is of opinion that at least the part of Ruanda bordering on Ankole consists of granite, contrary to Salée's representation (see fig. 17).

A few remarks of Salée may be quoted here [1]:

"Bien que son sommet soit à 1900 M. d'altitude, le mont Gatsibu est dominé à l'Ouest et au Nord-Ouest par des hauteurs plus élevées encore: Mont Gishikiri (2021 m.), Honga (2002 m.), Muyumbu (2247 m.). Kabimbiri (2272 m.). En somme le mont Gatsibu n'est qu'un des contreforts les plus orientaux du massif montagneux du Rukiga qui règne à la

---

[1] Achille Salée — "Constitution géologique du Ruanda Oriental", published in "Mémoires de l'Institut Géologique de l'Université de Louvain (Tome V, fascicule II, pp. 63 et seq.)"

frontière de l'Uganda et du Ruanda et dont les falaises occidentales vont tomber brusquement dans le Graben latéral du Mufumbiro. Du mont Gatsibu, on aperçoit la file des contreforts orientée Sud-Est Nord-Ouest; tandis qu'à leur pied vers le N-E, l'est et le S-E se déploie un pays mollement ondulé qui se maintient généralement entre 1400 et 1500 mètres d'altitude, et d'où émergent quelques pitons. Ce pays se relève vers l'Est jusqu'à une altitude de 1700 mètres pour former un rempart abrupt, bordant les plaines marécageuses de la rivière Kagera (1300 m.)".

This is a beautiful description of a typical arena. Referring to the descent from Mt Gatsibu to the arena, he continues:

"Constituant la route, on longe de véritables falaises constituées de quartzites blancs un peu micacés, alternant avec de fines bandes de schistes lustrés de direction N. 25° O. accusant un pendage O.—70°. . . . il s'agit ici d'un facies métamorphique des couches inférieures de notre Système de l'Urundi: Schistes et quartzites intercalés de la Lufironza".

Probably he refers here to the silicified, tourmaliniferous and metamorphic sericite phyllites, for in outward appearance the quartzites do not seem to differ from those of Ankole:

"des blocs de quartzite très riches en tourmaline noire ou foncée dont les cristaux prismatiques forment parfois de véritables masses, lardées de quelques lignes de quartz jaunâtres."

Regarding the original state of the lustrous and micaceous schists he says on page 65 of his book:

"Il s'agit bien d'un facies métamorphique des couches U 1, dû au voisinage des venues granitiques; car cinq cents mètres plus loin, les schistes ne sont plus lustrées, ni micacées. Leur couleur est gris bleu".

These are, therefore, the upper or lower sericite phyllites, the lowest horizon of the lowest division of the Karagwe-Ankolian system.

According to Salée, also the Ndorwa granite is a "granite gneissique à biotite", as might be expected, seeing that most arenas are formed by this variety. Further, the W. and SW. arena walls consist of metamorphic phyllites, like the NW. (explored by Miss A. Ledeboer) and N. arena wall.

From fig. 17 it may be seen that granite outcrops are only abundant along the periphery of the arena and particularly occur on the spurs of the arena wall. Farther towards the centre, however, the granite rarely protrudes above the thick detrital cover.

According to the principles set forth in part 1, it may be expected that the surface of the detrital layer chiefly consists of coarse fragments of almost unweatherable rocks. Moreover, these fragments will remain at the surface.

In the Ndorwa arena (as on the north flank of the Ihunga) these coarse, unweatherable fragments are supplied by quartzites outcropping on the arena walls. Further, amidst the granite on many of the hills Salée discovered outcrops of metamorphic quartzite in situ; also Miss Ledeboer found the same and called these roof pendants "baked roof quartzite". The present writer made a careful study of this phenomenon in Wishikatwa and will refer to it in more detail in Part III.

In fact, besides outcrops of granite and quartzite, many fragments of the latter have been found by Salée in the Ndorwa arena. To quote his own words:

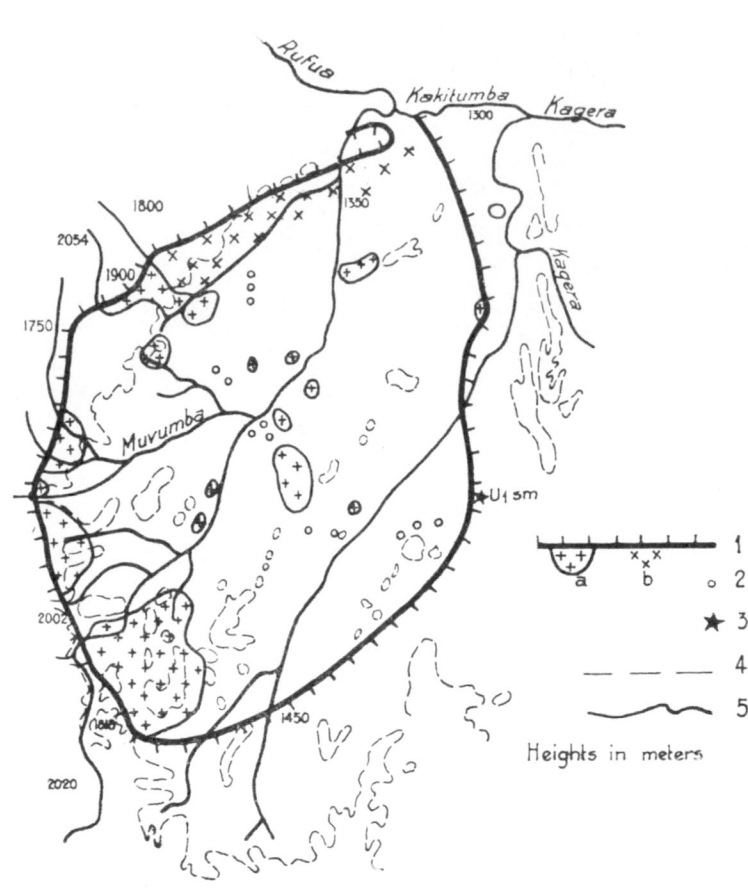

".... on commence à rencontrer une fine grenaille quartzitique [1]); puis les cailloux de quartzite montrent des tourmalines nombreuses ....".

or:

".... le sentier est parsemé de petits cailloux de quartzite .... nous rencontrons des cailloux de quartzite micacé avec tourmaline..."

The presence of coarse fragments of quartzite at the surface, however, is no indication that the bedrock in the immediate vicinity is formed by this kind of rock.

Fragments of schist, which are very soon weathered and pulverized, would have given much more positive indication as to the nature of the bedrock, but Salée found no fragments of schist; the first fragment was encountered far to the S.E. near the Kikoma (cf. fig. 17).

Fig. 17   The Ndorwa arena 1/750000.
1   =   Boundary of granite according to the conception of the present writer.
1a   =   Granite outcrops mapped by Salée.
1b   =   Granite after Combe.
2   =   Finds of quartzite fragments after Salée.
3   =   Find of schist after Salée (U1 sm).
4   =   1500 metre contour.
5   =   Rivers.

The present author, therefore, is very much inclined to map exactly what has been found, namely granite with roof pendants consisting of quartzite.

---

1)   Certain indication for granite, when amidst the Lower Division.

The fact that many fragments of quartzite are observed all over the arena floor may be considered as a good illustration of the influence of flood sheet.

CHAPTER 5

CORRELATION

One fact is certain, and that is that the Karagwe-Ankolian system is in all probability one without fossils. Presumably there is no unconformity or disconformity in the system, although during the deposition of the sandstone horizon there may have been a period of denudation. For the Upper Division the degree of load-metamorphism lies undoubtedly in the region of the epizone of metamorphism, and for the Lower Division probably on the border of the epi- and the meso-zones (garnet phyllites). It is distinguished from the Katanga system by the absence of calcareous rocks and dolomites, the Karagwe-Ankolian system being of an eminently detrital character. In composition and also in degree of metamorphism it resembles more or less the Witwatersrand system. The absence of fossils cannot be ascribed to load metamorphism, nor to lack of geological knowledge of this system, for it has been thoroughly explored by numerous geologists and prospectors, especially in trenches, tunnels and winzes. The lack of fossils is, therefore, a positive factor in determining the geological age. It is younger than the gneisses of the African "Primärformation", the "Afrizidische Grundgebirge", which Gregory calls Eozoic. It is undoubtedly older than the Karoo, which is found in Tanganyika in the form of continental deposits with fresh water shells and Glossopteris flora; which is found at Mombasa as the Mazeras sandstone with wind-worn pebbles and Permian fossils as also fossil wood, and which, finally, is encountered in the Congo with Triassic fossils (e.g. Lualaba-Lubilashe) and recently also with coal seams and Ecca fossils. The Karoo is also seen in an almost undisturbed position in Uganda, for instance at Entebbe, with Ecca fossils, and there is not a single reason to suppose that the Karoo might occur elsewhere without fossils. As a matter of fact most geologists no longer correlate the big non-fossiliferous systems with young palaeozoic sediments and few would correlate, for instance, the tillite at the base of the Kundelungu layers (or the tillite in van Doorninck's series IV of the Katanga system) with the Dwyca, especially since real Ecca layers have been discovered on the shore of Lake Tanganyika near Albertsville.

In the Devonian period Central Africa was probably land, whilst in the Sahara folded spirifer-sandstone and Devonian fauna with mediterranean facies are found. In Cape Colony there is the south coast of the Devonian continent and here the Bokkeveld series with Coblentzian fossils and the

Witteberg series with remains of Devonian plants were deposited. Why in Central Africa a non-fossiliferous facies of the Devonian should occur is not clear. On the contrary, the fact that no fossils occur is no reason to establish arbitrary correlations. Gregory, too, considers the absence of fossils a positive factor. It is less certain, however, whether the non-fossiliferous systems — which for a part at least are undoubtedly composed of continental depostits — do not belong to the *older* palaeozoicum. It cannot be absolutely denied, but neither is it at all probable, unless it were taken for granted that the limestones and dolomites of the Katanga system were deposited in entirely closed-in basins inaccessible to living organisms. Recently, however, there have been discovered in some of these limestones what are thought to be algae, and if this is the case it would mean that life had indeed already entered the basins.

The probability is that here we have to do with an equivalent of the North American or Scandinavian Algoncium.

The occurrence of tillites has also been exploited by various writers for correlation, but a study of the literature on this subject will show that so far not a single system has been found in which tillite does not occur. The Karagwe-Ankolian system forms no exception to this rule, for Wayland believes to have found tillite also in upper Karagwe-Ankolian sediments in Bunyoro. A tillite, therefore, does not form any fundamental distinction, and in this respect the writer is of opinion that he cannot do better than refer to what Van Doorninck has written about tillites and correlation ("De Lufilische Plooiïng").

So long as a thorough geological survey of Africa leaves such large gaps as exist at present, one will have to be content with the general composition, the folding and the degree of metamorphism. Some light, however, is shed on the problem by the now fairly generally accepted fact that the Katanga system is of about the same age as the Transvaal system. But as regards the age of the Karagwe-Ankolian system this is still an undecided question, though here, too, there seems to be a ray of light in the darkness, since, according to verbal information, detrital cassiterite has been discovered by Mr. Groves of the Uganda Geological Survey in the Bukoba sandstone. There is no doubt about this cassiterite having been derived from the Karagwe-Ankolian system, so that the system to which the Bukoba sandstone belongs must be younger than the Karagwe-Ankolian system. And as a matter of fact this already seemed to be most probable, for the system of the Bukoba sandstone, named by Salée the "système de la Lumpungu" and by others the "Tanganyika system", extends along the Victoria Nyanza in a narrow strip of land towards the south; beginning slightly to the south of Masaka, in Karagwe it leaves Lake Victoria near the Emin Pasha Gulf and continues parallel to the Lumpungu and the upper Malagarasi to Lake Tanganyika. This belt is nowhere very broad. Salée reports the following rocks in upward succession:

1. Grès feldspathiques et psammites: Grès de Goma.
2. Schistes argilleux verdâtres, phtanites, psammites: Couches de Muninya.
3. Calcaires gris-bleu avec cherts, en partie dolomitisés: Calcaires de la Musasa, into which rocks numerous basic dykes intrude. It is certain that in this region also other systems occur, and maybe also the Karoo, but it is highly improbable that the whole complex of gently folded sediments outcropping in the Malagarasi area has been deposited in that period, as some would have us believe[1]) "in the absence of definite palaeontological evidence to the contrary". *It is not by any means to be expected that non-fossiliferous systems could furnish the "palaeontological evidence to the contrary".* Rather, positive evidence that the "Tanganyika system" contains everywhere Karoo fossils has yet to be found. All that can be said at present is that the Bukoba sandstones, which occur together with calcareous sediments, are most probably younger than the Karagwe-Ankolian system.

Also Salée took the Tanganyika system ("système de la Lumpungu") to be younger than the Karagwe-Ankolian system. Both these systems are undoubtedly separated by a fault. The "système de la Lumpungu" is very much disturbed by faults or flexures, but on the whole shows only gentle folding. Granites are absent.

The author saw these sediments along the road from Kyaka to Bukoba, and the first impression was that they are less metamorphic than those of the Karagwe-Ankolian system. Thanks to Mr. Groves' discovery their relation to the latter system has now been definitely established. The writer, therefore, is in favour of the distinction made by Behrend, viz.: between the arenaceous — non-calcareous and the calcareous — non-fossiliferous systems, such as the Witwatersrand and Transvaal systems, or the Karagwe-Ankolian and the Katanga systems (according to Van Doorninck), in which latter system the writer would incorporate, for the present, the Tanganyika system.

When proceeding to make correlations one should take the well-known sections of these systems in SW. Ankole and the Haut-Katanga as specimens. It will then probably be found that the Karagwe-Ankolian system is much more extensive than it now appears to be, mainly owing to the confusing number of names. It may be that the Mafingi Mountains are built up by a system very similar to that of Karagwe-Ankole. This system may even occur on the farther side of Lake Edward in the Belgian Congo, as also **E**. of the Rungwe volcano. It is also possible that the Portuguese Umkondo system is to be correlated with it. The Umkondo system is described by C. Freire de Andrade (15th International Geological Congress, Pretoria) as follows: "The rocks of this system are chiefly sandstones, quartzites, argillaceous schists and phyllites. The last two form the base of the system and have purple and rose colours. Dolerite dykes and diorites are intrusive into the rocks of this system".

---

1) cf. E. O. Teale, 15 th Int. Geol. Congress, Pretoria.

Without desiring to correlate these rocks exactly, the author is inclined to assume that after the folding of the "Afrizidische Primärformation" principally detrital, non-calcareous sediments were deposited in perhaps separate basins almost simultaneously, which sediments are therefore comparable to the Algoncian sediments of Canada, Fenno-Scandia and Scotland. Further he is inclined to presume that, just as the Permian on the hercynian arch was deposited in separate basins and slightly folded (as in the Vosges), so here too a folding has taken place, though of much greater intensity (corresponding to the enormous orogenesis which preceded it), whilst acid magma likewise ascended. After this folding and subsequent denudation, presumably in a new phase the deposition and folding of calcareous sediments followed, as in the Katanga and Transvaal systems. In the Cambrian, Silurian and Devonian periods Central Africa may have been at rest and not until the carboniferous period was a continental series of sediments deposited, followed by faulting, but not, at least in Central Africa, by folding. It is according to this simple scheme, for which the author believes strong arguments to exist, that the appended tentative map (No 2) of Central Africa has been compiled.

CHAPTER 6

DESCRIPTION OF THE GRANITES — ABNORMAL TEXTURES — BASIC DYKES

A brief description of the granites is not out of place here.

The author would much rather leave it to some more capable and specialised petrographer to describe these granites, but as such a description does not exist, or at any rate has not yet been published, he feels obliged to make at least an attempt to give a brief description.

These granites indead yield very interesting features making a thorough investigation well worth while, in spite of the apparent monotony; owing to the majority of the granite exposures consisting of biotite-muscovite granite these intrusions do indeed seem to be of a character of but very little complexity.

In this granite biotite is the only femic component, the granite being built up further by muscovite, feldspar and quartz. Just as in all the Karagwe-Ankole granites, potash-feldspar occurs here as microcline, orthoclase never having been established definitely. Microcline-microperthite naturally occurs, but this is not very abundant. The sodium-calcium series of the plagioclases is represented almost everywhere. Both oligoclase and albite occur, the former generally in idiomorphic laths or as stout tabular crystals often embraced by microcline, and the latter as sometimes rather large crystals. Usually albite is decidedly subordinate to microcline, but in some places it increases in quantity and even reaches such proportions as to predominate entirely over microcline. The albite shows the twin lamellae typical of the albite law, whilst commonly

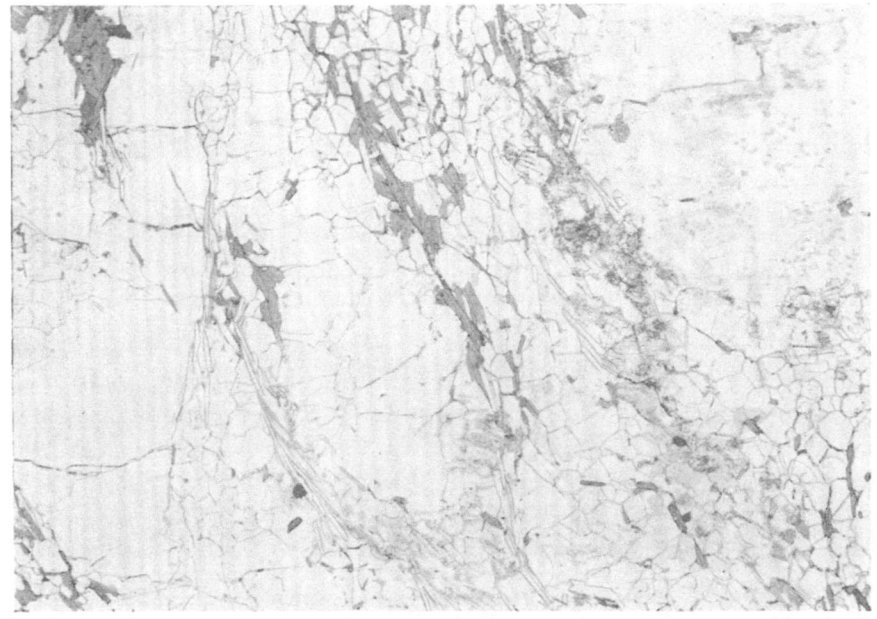

Photo 18a. (118n, magn. 12x, ordinary light) Gneissoid biotite granite N. flank of Ihunga.

Photo 18b. same as 4 a under crossed nicols.

plagioclase does not show this twinning, except in the more gneissoid varieties, where twin lamellae are encountered both of the albite law and of the pericline law.

As a rule many small mica flakes are included in the plagioclase cores. Biotite, which is always intergrown with muscovite, is highly pleochroic, the colour varying between dark brown and brownish white, whilst numerous zircone crystals produce the well-known halos. These zircone crystals, together with rounded apatite grains, form the accessory minerals. Quartz is present in abundance.

The biotite-granite, though in some places developed as a lepidoblastic granite-gneiss (e.g. between the eastern outlet of Lake Karenge and Katerero village), has generally a lenticular, more or less gneissoid texture.

This texture is due to the presence of coarse, badly shaped crystals of microline, in fluidally arranged mosaic of small, anhedral grains of quartz and feldspar, with small, idiomorphic scales of biotite and muscovite (cf. photo 19, U. 76).

Near the contact with the sediments these coarse microcline crystals sometimes attain such dimensions that one might speak of a porphyritic development. However, the contact is also often formed by a very fine-grained variety, which has a streaky appearance, owing to the presence of nodules of biotite and muscovite scales (e.g. east of the Omutarraz).

Photo 19. (76, magn. 17x, crossed nicols) taken from N. flank of Ruamheji. Shows myrmekite along edge of coarse microcline crystals.

These streaky, gneissoid varieties prevail near the contact with the sediments, but at a greater distance from the latter these varieties merge into more massive and coarse-grained types. As, however, far from the sediments the granite is mostly hidden by a thick detrital layer and exposures particularly occur near the margin of the batholith, the gneissoid variety is the ordinary type of granite generally encountered.

Whilst in the granite of Rushenyi, Lugalama, Wishikatwa, etc., the biotite is always absolutely fresh and does not show any signs of bending, this is not the case with the Rubanda granite. In fact this latter granite shows evidences of cataclasis, in contrast with the granites in Ankole.

This granite of the Rubanda boss is also principally a biotite granite. Its appearance is streaky, layers of pink feldspar alternating with small layers and lenses of quartz, while often the feldspar crystals are embedded in a fine-grained ground mass consisting of biotite, muscovite, feldspar and quartz.

Photo 20 gives evidence of crush and bending of crystals (mica), while recrys-

tallization is confined to the finer grains between the microcline crystals. The
latter crystals show a markedly undulatory extinction.

Deuteric processes, too, have taken part in the sealing of triturated zones, both in the Rubanda granite and in the Ankolian granites. Myrmekite is abundantly present in all these granites, and photo 21 suggests that a triturated zone between two microcline crystals has been sealed by the formation of myrmekite.

A part of the granite which obviously crystallized later than the biotite granite is the coarse-grained leucratic pegmatite granite. This type is found, for instance, in Wishikatwa and lacks every trace of Fe and Mg. It is

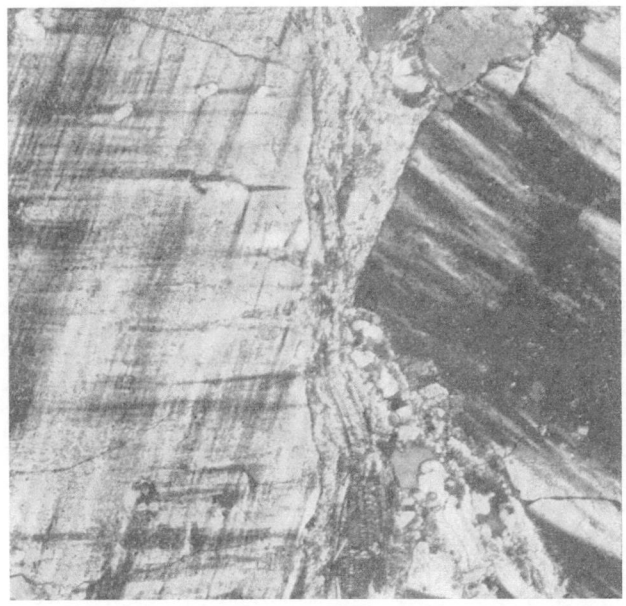

Photo 20. (No. 105, magn. 16x crossed nicols). Rubanda granite, showing triturated zone between two microcline crystals, with bent mica.

formed by the same constituents which build up the biotite granite, only

Photo 21. (93c, magn. 37x, crossed nicols, N. flank of Ihunga). Formation of myrmekite along a homogeneous microcline contact, resembling a sealed triturated zone.

biotite being absent. Sometimes it seems to constitute younger intrusions in the biotite granite (north of Ruamheji), but in eastern Wishikatwa it

Stheeman, Geology.

4

forms the core of a batholith. It is very interesting to observe that this younger leucocratic granite is not so much affected by stress as the biotite granite. Photo 22 is taken from a slide of the pegmatite granite.

Photo 22. (U. 58, magn. 16x, crossed nicols, Wishikatwa). Slide of pegmatite granite with albite (bottom left-hand corner) and microcline in which coarse scales of mica (top left-hand corner) and crystals of plagioclase with fresh rims. A veinlet of quartz runs diagonally through the picture.

A large part of the photo is occupied by microcline with vaguely defined twin lamellae; in the bottom left-hand corner parts of two albite crystals can just be seen. Under the microscope even this rock shows a somewhat abnormal texture, the quartz occurring in veinlets and lenses instead of in interstices. Besides coarse muscovite flakes also small, sericitized plagioclase laths are found in the microcline and these laths commonly have a fresh rim, which reminds one of myrmekite but in which mostly no radial quartz capillaries are to be detected. However, here and there one sometimes observes very fine capillaries (middle of right-hand edge), and this myrmekite, which consists of a rather acid plagioclase, penetrates along cracks into the potassium feldspar. Though this granite does not show a streaky or gneissoid texture, on the whole it gives the impression of being stressed with subsequent silicification and formation of deuteric minerals, e.g. myrmekite.

The femic pole of the granite is a true gneissose rock. In hand specimens, however, this rock could easily be taken for a sheared and crushed granite (photo 23).

Photo 23. (73, magn, 16x, crossed nicols, north of Mtunda.) Augite-amphibole granite. The amphibole (dark) is seen along the grain boundaries of feldspar and quartz grains, especially in middle and left edge of photo. Small (bright) grains of pyroxene with good cleavage are to be seen, especially along top edge.

It is built up by anhedral grains of quartz, microcline, albite and plagioclase, whilst myrmekite is also present in abundance in the microcline.

Further a few euhedral scales of biotite are to be seen. Pyroxene is present in rounded grains. Much amphibole occurs along the grain boundaries of all the other components, often embracing (in the section of the slide!) feldspar and quartz, thus forming sponge-like individuals. It is certainly not an alteration product of augite, as it only occasionally envelops the latter.

This amphibole is undoubtedly the last mineral in the sequence of crystallization, and the author surmises that it has been formed by metasomatic changes during the recrystallization of the stressed granite, from which this rock may have originated.

In both the biotite and the pegmatite granite basic dykes are found, particularly in the area north of Ihunga. Consequently these dykes are younger than the granites, though there may be differences in age between the various dykes.

Those which are considered to be older show a marked cleavage and a more advanced epidioritization.

These dykes are formed of fine-grained, equigranular, dark green rocks forming dykes in the granite which run almost parallel to the strike of the adjacent sediments and the foliation of the granite. They are sometimes rather long and of varying thickness, though seldom thinner than 1—2 metres, while they show a marked cleavage.

Two types may be distinguished, a hard variety of a dark green colour and showing little cleavage, and a less resistant variety of light green with pronounced cleavage.

The first, harder variety consists of amphibole, titanite and plagioclase. The sieve structure of the amphibole predominates and the amphibole is strongly pleochroic from yellowish-white to green or bluish-green. The maximum oblique extinction amounts to 24 degrees and the cleavage, birefringence and index of refraction are typical of amphibole. Plagioclase is present in small grains, often with twin lamellae, and has an index of refraction corresponding to that of oligoclase.

The second, softer variety, which is considered to be older, has undergone a much more pronounced change. It consists of the same amphibole but does not contain any plagioclase, only quartz being present.

In this variety large pyroxene crystals are found with the typical oblique extinction and cleavage. They are quite colourless and lack pleochroism, whilst the high polarisation colour proves that it is one of the monoclinic members of the pyroxene group.

No titanite has been found, but a strongly pleochroic biotite was observed with a colour varying from yellow to dark brown.

Ophitic texture is absent and whatever name is given to these rocks (Combe calls them dolerites) it is certain that they have been greatly altered by metasomatic processes. The author believes that these alterations occurred during the final phase of stress, which also produced cataclasis in some contact rocks [1]).

At the northern foot of the Ihunga stanniferous pegmatite-dykes cut the older basic dykes. Consequently these stanniferous pegmatites, which will be discussed in part IV A, are still younger than some of the "dolerites".

In the next part the reader will see that these granites have been intruded and have consolidated during the orogenesis. The abnormal textures might, therefore, be taken to be related to the folding. We may leave the further study of this problem to a more able petrologist. It must once more be emphasized that nothing has actually been observed in regard to real cataclasis (apart from the Rubanda granite). These granites bear no resemblance to the protogine of the Alps, and chlorite and epidote, for instance, are totally absent.

---

[1]) Still other basic rocks occur in the granite, particularly as marginal intrusions. These rocks are in an advanced stage of weathering. Combe has termed them "chalcedonic-tremolite rocks" and "diopside-plagioclase rocks". These are found S. of the Nerionza and S. of the Nya Bagaritshi, W. of the Ruitonbero and E. thereof in the southern branch valleys of the Rufua. The author has noted their existence but paid no further attention to them.

PART III

TECTONICS

VARIOUS PHASES OF FOLDING — NATURE OF CONTACT — MANNER OF INTRUSION

The Karagwe-Ankolian system forms one continuous outcrop in Karagwe, Ruanda-Urundi and SW. Uganda. It consists of a series of folds, the strike of which, from NNE-SSW near Lake Tanganyika, changes in SW. Uganda to SE-NW (see fig. 18). Three main elements are to be distinguished:

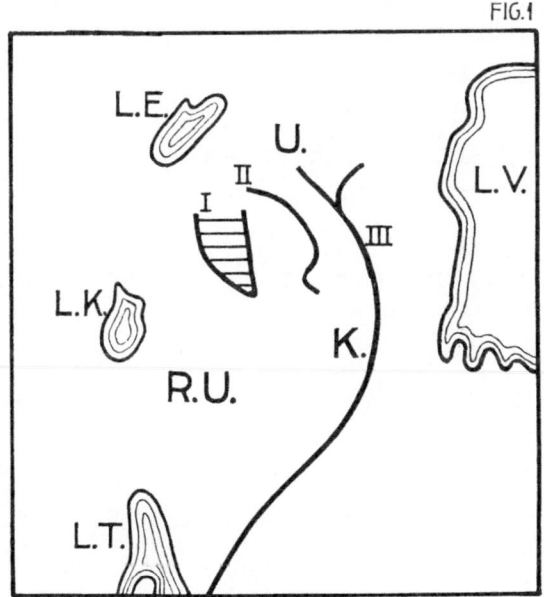

FIG.1

I. the Rukiga synclinorium dying out SW of Gatsibu and forming the Rukiga Mts., which extend from Ruanda into Kigezi;

II. the Kavungo syncline dying out SE of Gatsibu;

III. the Karagwe syncline, which in a sense is the backbone of the system and which (according to Salée) extends from Lake Tanganyika into SW. Uganda (see fig. 18).

In the intermediate anticlines large granite bodies are exposed (cf. map. 3).

Judging from the few scattered outcrops occurring in the west and middle of Uganda (cf. map 2), it seems that in those parts the strike turns again in the original NE. direction. Thus the strike describes a curve resembling, according to Wayland, a mark of interrogation drawn the wrong way round.

Fig. 18. L E = Lake Edward
L K = Lake Kivu
L T = Lake Tanganyika
L V = Lake Victoria
K = Karagwe
U = Uganda
R U = Ruanda-Urundi

The above mentioned synclines reveal outcrops of the Middle and Upper Divisions of the system. Salée has mapped the region around Kigali and Urundi as consisting of rocks of the Lower Division. It is quite possible, however, that the synclines occurring in Uganda extend much farther to the SSW, without the typical sandstone horizon occurring in the core of the synclines, but no proof of this is to be found in the data so far published. It would no doubt be worth while to apply to Urundi the stratigraphical knowledge obtained in Uganda.

As already remarked, granite bodies occur in the anticlines. As a matter of fact these bodies practically fill the anticlines, but a study of map 3 or map 4 will show that the granite in one anticline is divided by transverse syclines into several parts, each of which in itself, however, is large enough to deserve to be called a batholith. As indicated in Part I, the arenas described there coincide approximately with these granite areas.

The occurrence of these transverse synclines is a most remarkable phenomenon. Whereas the orogenetic force that caused the folding of the Karagwe-Ankolian system undoubtedly acted in an E-W or SW-NE direction, in the anticlines one finds minor synclines with a decidedly SW-NE strike. This strike is practically parallel to that of the Karagwe-Ankolian outcrops in Bunyoro and Buganda, so that it may be assumed that the interplay of directions has resulted from the shape of the basin. Most probably it was only in a later phase of the orogenesis that the influence of the Bunyoro direction was felt in SW. Ankole, since it will be seen that in SW. Uganda the compressional force at first acted roughly in a W-E or a SW-NE direction and that it was not until later that a force made itself felt in a SE-NW direction.

At first glance one would assume from the maps that the detail structure of the synclines is not in conformity with the general tectonics. The transverse synclines which give rise to this impression consist mainly of rocks of the Lower Division, that have been steeply wedged-in between the various granite bodies. It is probable however, that these bodies form one whole, as it is most likely that as a result of advancing denudation the transverse synclines will gradually disappear, the various granite bodies, such as those of Ndorwa, Rushenyi, Lugalama, etc., then appearing as one whole.

Farther on well-founded arguments will be advanced also for this hypothesis, but it may be pointed out here right now that the Ndorwa and the Rushenyi granites are already connected one with the other: Combe's mapping has established the fact that the syncline separating these granites (the so-called Buramma syncline) does not extend to the Kavungo syncline. From a short visit to the Mpungu mountain [1]) the writer gained the impression that the Buramma syncline rises out of the granite, much like the bow of a ship rising out of the water.

First of all, however, it is better to turn attention to the form of the contact between granite and sediments.

In most cases the sediments dip away from the granite, as in the arenas of Ndorwa, Lugalama, Chitwe, Ruberogoto and Ibanda, of Ntungamo, Kyer-

---

1) the NE. end of the Buramma syncline.

wa and Rubanda. The same applies for a large part of the Rushenyi granite.
However, overtilting of the sediments in respect to the granite also occurs,
particularly where smaller synclines lie in the granite, and in every case this
overtilting is directed towards the granite
body with the smallest outcrop. For instance
the north flank of the Ruamheji syncline is
overtilted (cf. figure 19) in a direction away from
the Karenge or Kazara granite towards the
Ihunga granite (cf. map 4 and section 5). Further, from section 3, Part II, it will have

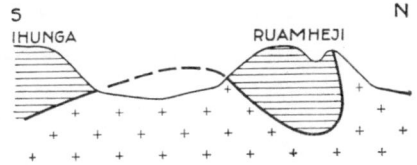

Fig. 19. Hatched = sediments; crosses
= granite.

been seen that the north flank of the Mushash-Kakanenne was also overtilted,
away from the Karenge granite and towards the Lugalama granite (cf.
map 4 and fig. 20). The eastern flank of the Chamiombu syncline (cf. fig. 21
and maps 4 and 5) is likewise overtilted towards the Lugalama granite.

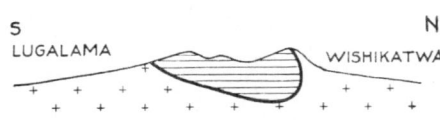

Fig. 20. Same indications as fig. 2.

It would seem, therefore, that a compression has taken place without any definite
direction, whereby the granite with the larg-
est outcrop has endeavoured to spread out
over a smaller dome (cf. figs 25 and 28).
Considering, however, that as a rule the sediments dip away from the granite bodies and thus the longer the latter are exposed to denudation the
larger the outcrops therein become, it may also be said that the granite dome,
which once reached a higher level, tried to spread out over a less elevated dome.

Very small and narrow anticlines
have played quite a passive rôle. Thus
the north flank of the Shonobutondo
anticline is overtilted towards the
Wishikatwa dome and away from the
Rushenyi arena (section 2) under the
influence of the enormous Rushenyi dome (cf. fig. 25).

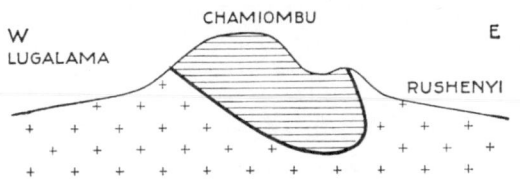

Fig. 21. Same indications as fig. 19.

In Kigezi only one granite body is exposed, the Rubanda granite lying in an
anticline. In that region, therefore, the Rubanda granite reaches the highest
level, assuming there are more bodies which have not been exposed by erosion.
Nevertheless, both on the Wramuyonyi and to the SW. of the Rubanda granite
the overtilting is directed towards the SW.

From sections made by Combe it further appears that between the Rushenyi and the Chitwe arenas (cf. map 3) the folds are overtilted towards the
former arena, which is the smaller of the two.

In this respect, therefore, the compressional force during the last orogenetic phase acted in no definite direction and on this base the section showing the structure of SW. Uganda (cf. fig. 28) has been drawn.

For the rest the sediments are practically in conformable contact with the
granite, as is sometimes to be seen on mountain slopes and in gullies, for in-

stance on the slopes of the Ihunga, the Ruamheji and Mtunda and also of the Chamiombu. This fact, however, is particularly to be deduced from the observation that in most cases, and in spite of great differences in level, practically the same stratigraphic horizon forms the contact with the granite, so that it may safely be assumed that this will also be the case at lower levels. In some parts this stratigraphic horizon is formed by the lower sericite phyllites, and in other parts by the upper sericite phyllites, but generally it is roughly the boundary quartzite, as shown already in Part II.

This horizon follows closely the boundary of the granite, the direction of strike sometimes sweeping as much as 360° around the granite, as very clearly shown on map 4 (the Lugalama arena). Consequently, above the culminations of the granite domes the whole system must have been very flat, almost horizontal.

In the secondary synclines between the various granite domes the Middle and Upper Divisions are absent, except between the Kyerwa and the Ibanda granites. This is undoubtedly a result of erosion. Both map 3 and map 4 reveal the steep plunge of the axis of the transverse synclines towards the major synclines; see for instance the syncline between Kabezi and Lutobo and that of Kaina. The axes of these synclines, therefore, were more or less parallel to the general trend of the now eroded sedimentary roof of the anticlines, in which anticlines only the lowest parts of the transverse synclines have now been left.

Undoubtedly these transverse synclines have been formed in consequence of a longitudinal compression, and as a result also the higher divisions were more or less folded, though probably in this enormous belt, consisting of plastic argillaceous and resistant arenaceous rocks, the effect of this folding gradually diminished upwards.

This phenomenon is demonstrated in the Ihunga, the Ihunga quartzite being much less steeply tilted than the lower horizons.

Indeed, the absence of the Middle and Upper Divisions is not everywhere to be explained by erosion, as for instance immediately west of Kaina or in the arena between Lutobo, Kabezi and Kamwezi (cf. maps 3 and 4). Although in these places secondary synclines split off, as it were, from the first Rukiga syncline, the Ihunga quartzite bends out but very little. From these observations the writer has evolved the following hypothesis:

1. In consequence of longitudinal conpression the *roof* of the batholiths has been folded transversely.
2. As a result particularly the more plastic lower horizons of the Karagwe-Ankolian system became wedged in the granite.
3. The axes of these secondary synclines ran parallel to the anticlinal arch of the roof.
4. Since the lower horizons were more steeply folded than the higher divisions, one would expect to find rather great differences in thickness in

the lower horizons, and as a matter of fact this is occasionally observed.

Thus there exists in SW. Uganda a superposition of two orogenetic phases with different directions of the respective compressional forces. The resulting interplay of directions is very interesting, though in the beginning also very puzzling.

The great importance in fully solving the problem of the tectonics of the Katuba valley lay in elucidating the effect of the two phases mentioned (cf. map 5). There is no doubt that the eastern flank of the Chamiombu-Nyi-hanga syncline once extended northward as far as the Ruitonbero, and a subsequent N-S compression caused a lip-shaped bulging of this flank in an eastward direction, thereby giving rise to the numerous faults east of hill 36. In the meantime, however, the compression continued and as a result of fracturing in a direction perpendicular to the axis the sediments were thrust up against the former northern extension of the Chamiombu syncline (fig. 22).

From hill 29 towards the west the throw of the fault increases, decreasing again near the Mush-ash. Between the Mushash and the Ka-vusanammi-East there is a badly disturb-ed north flank, in which the boundary quartzite is absent and in which the E-

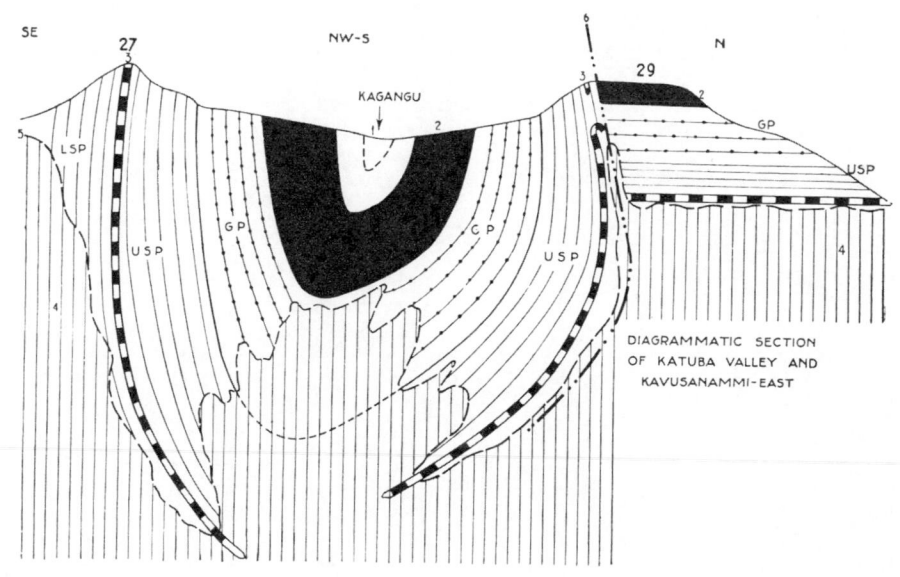

Fig. 22. 1 — Phyllites of Middle Division. 4 — Presumable position of granite.
2 — Ihunga quartzite. 5 — Presumable contact.
3 — Boundary quartzite. 6 — Plane of dislocation.

W strike forms a sharp contrast with the N-S strike of the roof pendants (consisting of quartzite with or without metamorphosed phyllites) lying in the granite immediately north of this north flank. Traces of the enormous pressure are still to be seen in the exceptional mineralization of the upthrust zone, in which coarse staurolites and very large cyanite crystals frequently occur.

The Kavunga syncline likewise formed two spurs, one of which — the Nabusov syncline with its axis plunging eastward — is still fairly intact; of the other syncline, the north Rushenyi syncline, only some remnants have been left, for instance east of Lubale (cf. map 4 and fig. 25). The northern flank of the North Rushenyi syncline, however, is also still fairly intact (cf. map 5 and fig. 25); this extends from the foot of the Nyihanga via hills 38, 41 and 42 to the east. Between the Nabusov, the North Rushenyi and the Katuba syncline there is now a narrow anticline [1]) in which granite is exposed.

This interplay of directions ceases in the west. In the middle of Kigezi neither a true E-W phase nor a true S-N phase has predominated. The area between Nya Lusanje and Muko (cf. maps 3 and 6) is already beyond the turning point, the orogenetic force apparantly having acted in a direction more or less NE-SW, and in fact north of Nya Lusanje there is a gradual transition from a more E-W to a N-S trend.

This change of direction takes place north of the Ruamheji syncline in the syncline of the Shumba-Nya Bagaritsyi (cf. map 4), of which only two roof pendants have been left, consisting of two quartzites dipping towards each other and accompanied by remnants of phyllite. Both flanks swing round in a more N-S direction. In fact remnants of sediments with a true N-S strike are found in the granite east of the Shumba-Nya Bagaritshi syncline (cf. map 4).

As already observed, this superposition of two different directions of folding caused a considerable dislocation, and this dislocation is characterised by the occurrence of a mineral, viz. cyanite, which otherwise is entirely unknown in the Karagwe-Ankolian system. Moreover it has to be pointed out that elsewhere staurolite crystals have never been found to occur in sericite phyllites [2]).

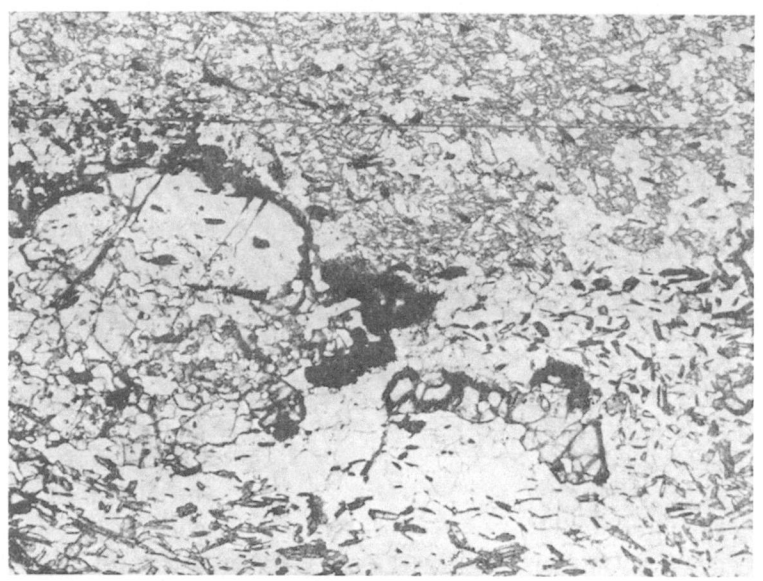

Photo 24. (U 37, magn. 16x, ordinary light). Slide of cyanitiferous, metamorphosed garnet phyllites in Chamiombu syncline. In middle garnet, top edge staurolite displaying typical sieve structure, bottom part small crystals of cyanite; all intergrown with quartz. Mica almost absent.

Cyanite, however, also occurs in another spot, in metamorphic garnet phyllites between hill 34 and hill 35 in the Katuba valley, and there too it is met with in strongly compressed rocks forming the nucleus of an overtilted syncline (cf. photo 24; also fig. 21 and section 1, Part II).

Inversely, therefore, it may be deduced from the occurrence of cyanite

1) The southern part of this (cf. map 4) is called the Shonobutondo anticline (cf. fig. 8).

2) They do indeed occur in the higher horizons of the system, but only as irregular grains with pronounced sieve structure (cf. photo 24, in which part of *one* crystal is shown), and never in the form of euhedral-shaped crystals found in the above mentioned zone.

that a rock in which this mineral is found has been subjected to considerable pressure, at least in Ankole. Under certain conditions also staurolite may be an indication of strong pressure. Now the occurrence of cyanite is not confined solely to the reverse fault discussed above, for it occurs not only in the zone from the Kavusanammi-East to the Mushash but also in the sericite phyllites on the Ruacherenze and the Kakanenne. It again appears in the north flank of the Ihunga, but not in the south flank, whilst the same may be said in regard to the Ruamheji syncline, where cyanite is met with in the overtilted north flank but not in the gently dipping south flank. It is seen, therefore, that the north flanks have been subjected to pressure more than the south flanks, and also that the cyanite occurs in a W-E zone, first formed by the north flanks of some synclines but subsequently extending in a dislocation.

Furthermore, in mapping the Katuba valley it has been found that the longitudinal compression which resulted in this dislocation has brought about deformations in synclines with a S-N strike (the first Rukiga syncline and the Chamiombu syncline). In addition also the SW. flank of the Kavungo syncline has been bent outwards, and these lateral spurs of the Chamiombu and the the Kavungo synclines are found to mesh one with the other.

The author believes that in the beginning the compressional force acted mainly in a W-E direction, with a northerly component in Kigezi and a southerly component in Bunyoro. Orogenesis proceeded without complications until the whole basin of the Karagwe-Ankolian system had become a fairly rigid mass. It was not until then that the longitudinal compression was set up in consequence of the advancing movement of Bunyoro and the resistance offered in SW. Uganda and Ruanda. In this phase also the transversal secondary synclines were formed and as a result also the biggest granite cupolas became more or less fan-shaped. According to this hypothesis the longitudinal compressional force originated in Bunyoro in the NW and consequently the dislocation of the Kavussanammi would have to be regarded as a sort of underthrust fault [1]), the occurrence of which is uncommon.

Another reverse fault occurs in the Buramma syncline. In the north flank of this syncline the garnet quartzite (cf. section 6, Part II) is badly disturbed; for large distances it is entirely absent, but even there on the boundary of sericite and garnet phyllites contorted lenticular fragments of this quartzite may be found. This is not to be ascribed to transverse faulting, since neither in the garnet quartzite in the south flank nor in the boundary quartzite in the north flank are there any traces of displacement by transverse faults.

The cause of the disturbed position of the garnet quartzite in the N. flank of the Buramma syncline must be sought in the presence of a longitudinal

---

1) Strictly speaking it is a normal fault, but it is essentially a product of compression and most probably the dip of the fault plane was formerly to the south. It may be that in consequence of the advance of compression the fault plane itself was overtilted together with the overlying upthrust sediments (cf. fig. 22).

reverse fault, along which the southern part has been thrust up against the northern part. The throw of this fault may be small (cf. map 4 and fig. 79 Part VB). Many signs of metamorphism are encountered along this fault.

Turning back to the dislocation of the Kavusanammi, it is seen that along this fault the granite has replaced the sediments; granite is found, for instance, between the Chikoba and the Mushash, and also between the Kazko and the Kavusanammi-West (cf. map 5). It appears to be beyond doubt, therefore, that the granite was still active during the second phase of the orogenesis. Furthermore it seems as if the granite is the base on which the Karagwe-Ankolian system has been deposited, and which has subsequently been folded simultaneously with the folding of that system, such is the regularity of the occurrence of granite in anticlines, which can only be explained satisfactorily if the intrusion of the granite is at least contemporaneous with the first phase of folding.

An active lifting of the sedimentary cover by the granite is out of the question, because of the nature of the tectonics being essentially one of horizontal compression. It was in consequence of this compression that the folds were formed, and as a result the granite apparently filled the regular, elongated spaces. The granite has undoubtedly flowed in passively and simultaneously with the folding, the nature of the contact with the sediments being intrusive. Formerly there was some doubt about this.

One of the remarkable phenomena is the absence of apophyses of the granite in the country rock, whilst very often one and the same stratigraphic horizon forms the contact with the granite.

There are three kinds of proofs that the granite is indeed intrusive, viz.:
1. actual proofs of replacement of sediments by granite;
2. evidences of contact metamorphism;
3. cross-cutting relations.

As to the proofs of the replacement of sediments by granite, these are furnished by the so-called injection schists, which will be described later in Part IVA. A. E. Speyer, who found these near Kabezi in pits at the foot of Buramma Ridge, described them as granite with remnants of schists; Combe discovered them west of the Nya Bagaritshi and reported on them as being large tongues of mica-schist included by granite. The present writer studied the occurrence of these injection schists SW. of Nya Makukuro at the foot of the Ihunga, where he noticed a lit-par-lit replacement by granitic material (cf. photos 28 and 35 Part IV A).

As regards contact metamorphism this may be said to have been established beyond doubt, though there are no striking contact rocks, nor well-defined and typical contact minerals. The horizon generally forming the contact, i.e. the sericite phyllites, is invariably converted in the vicinity of granite into more or less silicified, tourmaliniferous mica schists. The contact metamorphic aureole varies in thickness, averaging some 200 metres.

It seems that a quartzite has formed a considerable hindrance for extension of contact metamorphism, there being in many places no longer any trace of it on the farther side of the boundary quartzite [1]). Only *white* mica is present, whilst in addition to tourmaline and quartz also rutile and hematite occur.

By way of exception still other minerals are sometimes found in the sericite phyllites; the occurrence of cyanite and staurolite has already been described. It is, of course, quite unknown what rôle has been played by the tectonics and by the granite in their formation. Another mineral occurring in the sericite phyllites is biotite; it is found in the north flank of the Ruamheji syncline (photo 25, U 53, magn. 14 ×). In the vicinity of granite, the garnet phyllites and the phyllites of the Middle Division contain staurolite, which occurs as irregular, brown spots; it has evidently grown along the grain boundaries, resulting in a pronounced sieve structure (cf. photo 24). The tourmaline occurring, which is not often, is very fine and markedly

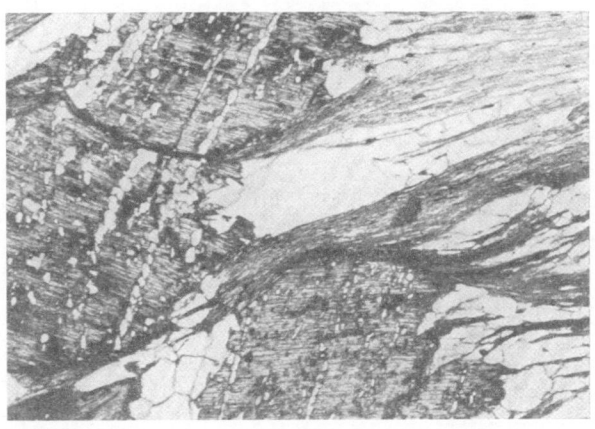

Photo 25. (U 53, magn. 14x, ordinary light). Eyes of biotite causing lenticular texture in quartzose mica schists. Peculiar veinlets of quartz cause pseudo-sieve structure in biotite.

pleochroic from impure white to dark blue. Only in one spot has cyanite been found in garnet phyllites (cf. photo 24), with simultaneous disappearance of white mica.

Finally the cross-cutting relations. These the writer studied for the first time in Wishikatwa, where he encountered quartzites lying in the granite sometimes with and sometimes without schists (cf. maps 4 and 5). Explorations between the Ruitonbero and the Nya Bagaritshi brought to light the occurrence of an anticline and a syncline in the granite. The course of the contact with the granite near the Ruitonbero could not be otherwise than as sketched in fig. 23. This was confirmed and at the same time explained by investigations in north Rushenyi: in the zone of tension, which is to be expected at the bottom of the synclines, there are fissures filled with granite (see fig. 24). There the granite has assimilated or removed by stoping a certain quantity of phyllites, whereby

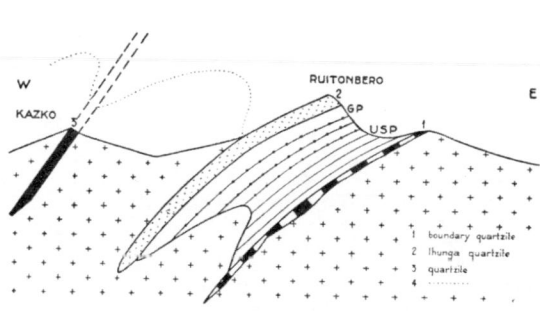

Fig. 25. Section through Ruitonbero and Kazko (cf. map 5). 4 = supposed former plane of contact now eroded away.

---

1) This is not the case on the north flank of the Katuba syncline.

the resistant and relatively insoluble quartzites have been spared. With deep erosion, therefore, signs of contact metamorphism will also be found in the core of the syncline and finally even granite will be met with in the synclines [1]). Map 4 shows examples of various stages of this process; where the author believed the granite to have replaced parts of the synclines this has

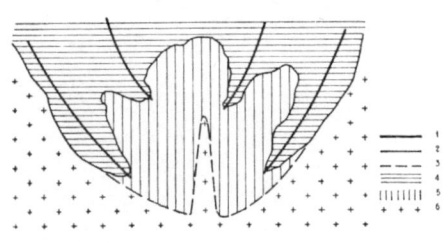

been indicated on the map. Fig. 25 illustrates the writer's conception of what once the roof of the granite looked like; it is a longitudinal section running partly parallel to the major strike (in the southern part) and partly forming an angle of roughly 45 degrees with that strike (in the north).

Fig. 24. Cross-cutting relations between granite and sediments in bottom part of a syncline.
1 = quartzite; 2 = present plane of contact; 3 = shape of initial wedge of granite in zone of tension; 4 = sediments; 5 = sediments replaced by granite; 6 = granite.

Reviewing the principal features of this granite, it is to be concluded that in all probability this is not a case of a bottomless intrusion, considering that:

1. the contact with the sediments is essentially conformable and is formed by one definite stratigraphic horizon;

2. underlying this horizon there is always granite, other rocks being unknown;

3. nowhere are apophyses of the granite to be found in the sediments;

4. founderings of the roof are extremely rare.

In the areas explored the writer found only three indications of foundering, the first being near hill 25 (cf. fig. 26) at a sharp bend of the strike. The same was seen west and east of Lutobo, where the strike likewise made a sharp bend. Owing to a normal faulting the boundary quartzite has partly disappeared and granite and sediment are in abnormal contact with each other.

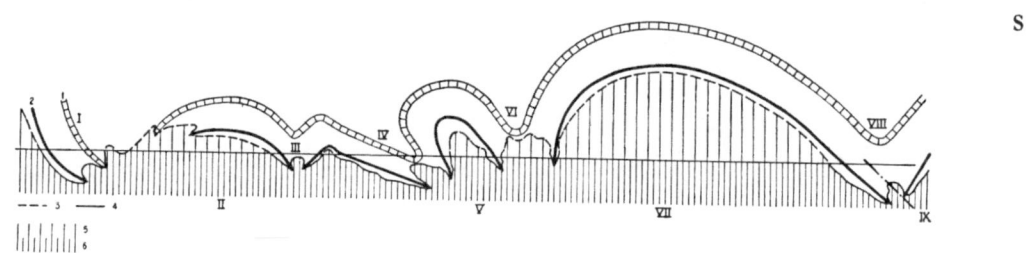

Fig. 25. 1 — Ihunga quartzite.
    2 — boundary quartzite.
    3 — hypothetical plane of contact.
    4 — actual contact.
    5 — granite eroded away.
    6 — solid granite.

    I — N.W. extension of Kavungo syncline in Kazara.
    II — Karenge or Kazara granite.
    III — Nabusov syncline.
    IV — Katuba syncline.
    V — Shonobutondo anticline.
    VI — North Rushenyi syncline.
    VII — Rushenyi granite.
    VIII — Buramma syncline.
    IX — Ndorwa granite.

---

1) In the Rushenyi granite, between hills 61 to 64, there are remnants of what is possibly quartzite, accompanied by injection schists, and this would point to the presence of another N-S syncline. This part has provisionally been mapped as consisting of granite, the boundaries and trend of the syncline being as yet unknow.

Quartz has been deposited along the fault plane.

A similar occurrence is to be seen near Kaina (see fig 27). A close examination of the first quartz reef, however, shows inclusions of remnants of quartzite. Slides have proved this beyond doubt, reveal-
ing at the same time that these quartzite remnants have been very little affected by metamorphism. From the Nobugamba to just beyond Kaina Hill no rem-
nants of quartzite are to be found.

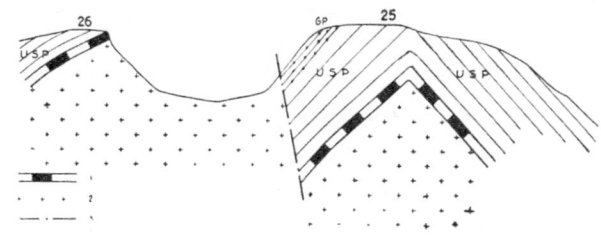

W. Fig. 26. 1 — boundary quartzite. E.
2 — granite.
3 — fault plane.

In the arena floor, some hundreds of metres in front of the arena wall, there
are very large, loose boulders in which quartzite is easily recognised, intersected and surrounded by vein quartz; they have undoubtedly been displaced from the arena wall.

The presence of quartzite fragments in the quartz reef is taken by the author to have been due to drag. It must be borne in mind that there is a great difference between the me-
tamorphosed quartz-
ite in the granites and this quartz reef; the latter is not to be regarded as a metamor-
phosed quartzite (cf. Part V A).

If it were to be as-
sumed that these gran-
ite bodies are bottom-

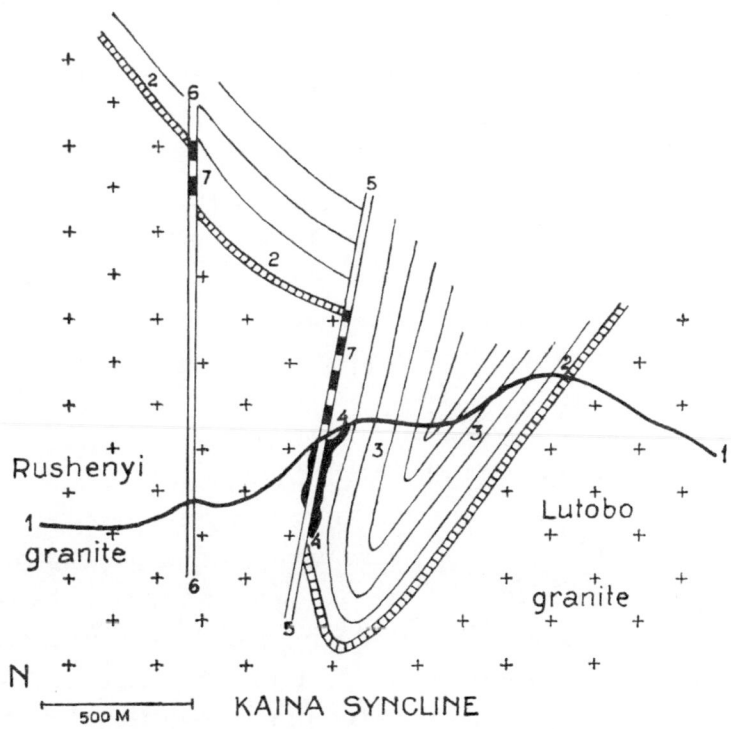

Fig. 27. 1 — present surface.   2 — boundary quartzite.
3 — upper sericite phyllites.   4 — metamorphic aureole.
5 — first quartz reef.   6 — second quartz reef.
7 — fragments of quartzite in quartz reef.

less intrusions it would be quite inexplicable why the roof is not much more
broken and how it is that foundering has not taken place on a larger scale.

The absence of apophyses also indicates that there was little faulting,
whilst the sediments were folded during the intrusion. One would be inclined,
therefore, to assume that the synclines were supported in some way or other.
Though, of course, it can only be conjectured, it would seem that during the

orogenesis the granite was injected along the base on which the Karagwe-Ankolian system has been deposited. This would also be a logical explanation for the absence of all rocks older than the upper sericite phyllites. This hypothesis has been worked out graphically in fig. 28.

In the last phase of the orogenesis the outer crust of the granite was probably already solidified, which would explain the lenticular and gneissoid texture of the granite, especially in its marginal portions. In addition to this, traces of crushing and granulation are to be found in some contact rocks. Traces of cataclasis, for instance, are found in stanniferous veins underlying

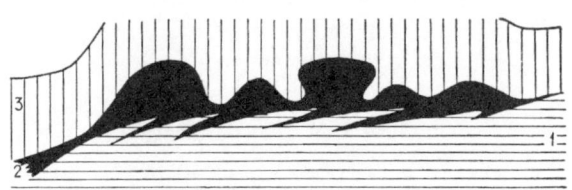

Fig. 28.   1 — Base on which Karagwe-Ankolian
                  system has been deposited.
             2 — Granite.
             3 — Sediments of Karagwe-Ankolian
                  system.

the boundary quartzite in the north flank of the Buramma syncline (cf. Part V C). Another instance is met with south of Rubanda (map 6), where the Rubanda granite outcrops in an anticline. Whilst the sediments dip away from this granite they do not sweep around it, and moreover the granite lies excentrically to the axis. Map 6 clearly shows the cross-cutting relations of this granite. South of Rubanda a most peculiar contact rock was discovered with a lenticular texture and yielding a detritus exactly similar to that of granite, viz. round, transparent quartz grains embedded in an argillaceous cement; feldspar, however, was absent. A microscopical examination (No 101) proved these schists to be strongly cataclastic, half of the rock being formed by lenticular eyes consisting of a pseudo-isotropic, structureless rock powder. These eyes include a few rounded fragments of quartz and very small flakes of sericite, streaks of fine-grained quartz and very fine-scaly mica separating the oblong eyes. Here the deformation has not been accompanied or followed by recrystallization. Apart from quartz and sericite or mica, one may observe pennine, pyrite and a few fine needles of tourmaline, which are strongly pleochroic, the colour varying from impure white to bluish black.

Finally, as an argument for the supposition that the granites in the various major anticlines are inter-related, it may be observed that in composition they are remarkably similar; a hand-specimen from the Ruberogoto arena does not differ from one taken from the Ndorwa or the Rushenyi arena. This similarity of composition of the granites also appears in map 4, and it is at the same time an argument for the supposition that already at the time that intrusion took place the magma was of a granitic composition and has changed but little since then.

PART IV

METAMORPHISM

A. AUTOMETAMORPHISM

CHAPTER 1

INTRODUCTION

Since the tin deposits show accumulations of those minerals usually called pneumatolitic, it is beyond all doubt that their genesis is closely related to the intrusion of the granitic magma [1]).

Moreover on the northern slope of the Ihunga coarse cassiterite is found in pegmatite dykes intrusive in the gneissoid biotite granite (cf. map 4). As these pegmatites also contain much tourmaline and coarse mica, both of which minerals are related to the numerous stanniferous veins in the sediments, the author wishes to describe the occurrences of these minerals in the granite.

This description serves a twofold purpose. In the first place it is interesting to observe that the sequence of deposition of the minerals crystallized in pegmatites and autometamorphic varieties of the granite is much similar to that appearing from an investigation of the veins in the sediments, and in the second place the ore-geologist wishes to know in what manner the solutions were formed by which stanniferous veins in the sediments have been deposited.

The pegmatites consist of potassium feldspar, coarse plates of white mica, coarse columns of black tourmaline and quartz. Cassiterite is present in coarse, euhedral crystals and beautiful penetration twins. It is a minor constituent, as the richest pegmatites carry a maximum of 0.06 % cassiterite by weight. The sequence of deposition of tourmaline, quartz and mica has not been studied with the aid of slides of the above-mentioned pegmatite dykes, it being impossible to prepare thin sections of these rocks, owing to the constituents of these deeply-weathered, coarse-grained rocks, which contain but little quartz, all falling apart.

Only from the pegmatite dyke of Katerero could good specimens be obtained. However, as far as could be discovered with the aid of a magnifying glass,

---

1) The biotite granite contains cassiterite, though it has only been observed in a single slide, in which it was present as very minute grains of a yellow colour.

Stheeman, Geology.

these pegmatites show exactly the same features as the pegmatitic rocks now to be described, from which good slides were obtainable.

These pegmatitic rocks occur near the boundary of the granite, marginal portions of which have been greatly changed by autometamorphic processes. These processes include the deposition of tourmaline and coarse mica and intensive silicification ultimately resulting in the formation of true "greisens".

All transitions occur from normal biotite granite to tourmaliniferous "greisens" and nowhere is a sharp contact between the two visible, proving that these latter rocks have not been formed by separate intrusions. Indeed texture and composition of these marginal, pegmatitic rocks are subject to considerable variations. Generally they consist of large, white microcline crystals with imbedded coarse, black, stout, prismatic, euhedral tourmalines, often from 2 to 5 cm. long. Under the microscope this tourmaline is strongly pleochroic and its colour varies between brownish white and dark bluish green or greenish blue. The quartz content varies, but increases towards the contact with the sediments. The quartz forms coarse, tubular veinlets in the feldspar but particularly occurs along the heterogeneous boundaries of the tourmaline and the feldspar. It has been clearly observed that the quartz has replaced the feldspar. The feldspar content decreases with the increasing quantity of quartz, whilst coarse muscovite is increasingly abundant where the silicification has been more intense. The tubular veinlets of quartz in the feldspar are often diamond-shaped in section, being solely bounded by the cleavage planes of the feldspar.

As regards texture it is impossible to describe the numerous transitional forms in all their varieties. Approaching the biotite granite one generally finds a finer-grained intergrowth of quartz and feldspar, the quartz veinlets becoming smaller in diameter and the muscovite diminishing considerably. Often no muscovite at all can be seen megascopically. The tourmaline prisms are more slender and sometimes form the well-known "suns" and radial aggregates and, being found in parallel orientations, they give rise to a nematoblastic texture. This variety changes directly into the gneissoid biotite granite, owing to the tourmaline disappearing entirely and simultaneously with the entry of biotite as a constituent. Of forty hand specimens only one was found to contain biotite and tourmaline together, and even in that case it was very difficult to get both these minerals in one slide (slide No 93 d).

The metasomatic changes which have resulted in the formation of the pegmatitic portions of the granite rich in tourmaline, quartz and mica were undoubtedly caused by emanations of the magma. Consequently these changes belong to the autometamorphic processes, and in particular to the additive autometamorphism, seeing that large quantities of elements have been added to the original granite which did not occur in it before; among these elements are fluor and borium, whilst also silica appears to have increased considerably.

In a narrower sense this type of autometamorphism is termed "pneumatolitic". This term, however, implies that the gas phase is considered to play an important part in this type of metamorphism. As the writer is not sure that this has actually been the case, the use of the term "pneumatolysis" will be avoided.

Photos 26a and b have been taken from slide No 70 of a hand specimen originating from the tourmaliniferous rocks occurring north of the Omutarraz (Wishikatwa) and representing the transitional zone between the totally silic-

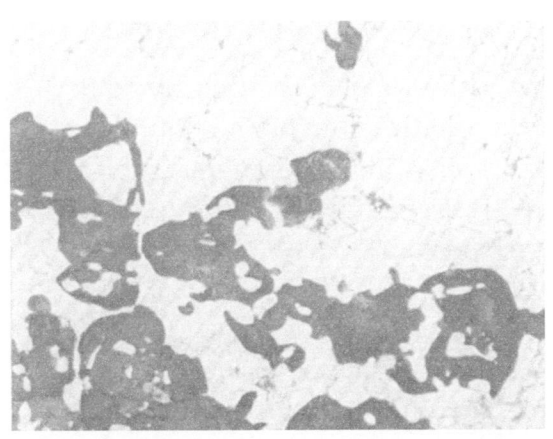

Photo 26a (No 70, magn. 14x, ordinary light). Showing tourmaline (black), quartz (clear) and feldspar (turbid).

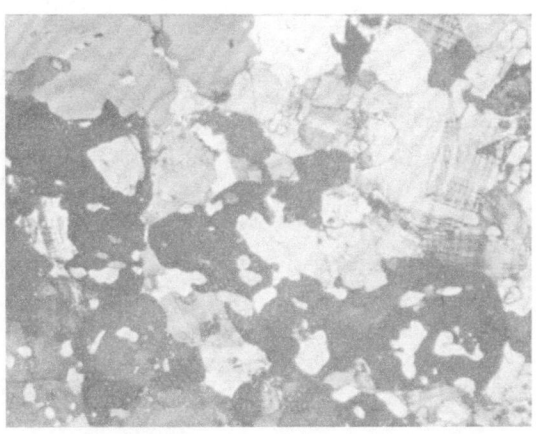

Photo 26b. Same as 1a under crossed nicols.

Photo 26c. Same slide under ordinary light.

Photo 26d. Same as 1c under crossed nicols, showing microcline, tourmaline and quartz (latter embracing small drops of feldspar).

ified "greisen" and the surrounding biotite granite. Photos 26a and b (magn. 14 x) clearly show that the tourmaline lies in a ground mass of small grains of microcline and quartz. Photos 26c and d, from the same slide, give good indications regarding the original granite. There is no doubt that the quartz was partly present already as a primary constituent, but another part must have been metasomatically deposited. This may be seen when examining the quartz crystal shown in the top left-hand corner of photos 26a and b, where minute round grains of microcline are seen enclosed in the quartz. The index of refraction of these grains is less than that of quartz and in most cases they show the typical twin lamellae of microcline. In the middle of photo 26d these microcline grains are easily seen in a bright quartz grain.

The tourmaline has been metasomatically deposited on the grain boundaries of microcline and quartz, enveloping both these minerals. Tourmaline is particularly found along the homogeneous feldspar boundaries, and apparently feldspar has been replaced by preference. It is probable that much of the enclosed microcline has subsequently been replaced by a later quartz, which after the deposition of the tourmaline might have penetrated along the heterogeneous microcline-tourmaline boundaries. This appears from many observations, a few of which may serve as examples.

Photos 26a and b show a triangular fragment of feldspar enveloped by tourmaline (half-way up the left-hand side). The tourmaline-microcline boundary runs on two sides parallel to the cleavage of the microcline. Two small droplets of quartz are present along the heterogeneous contact. In the bottom left-hand corner of 26c and d there is an inclusion in the tourmaline consisting of quartz with a remnant of microcline. The tourmaline on the left-hand side of photos 26a and b encloses principally quartz, in which the same round particles of microcline are to be observed as exist in the larger quartz grains. Each feldspar grain embraced by tourmaline appears to be a separate individual crystal, whilst several quartz grains enclosed by tourmaline prove to form one individual, since they are extinguished simultaneously. Several droplets of feldspar included in quartz, particularly when lying close together, also extinguish simultaneously; therefore the author takes them to be remnants that escaped silicification.

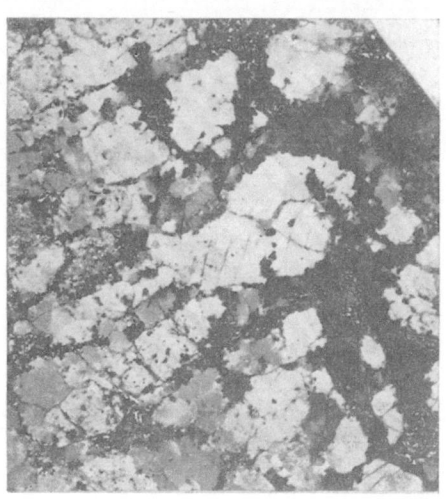

Photo 27. Autometamorphic variety of granite on northern slope of Ihunga (cf. map 4) — magn. 2x, polished surface, reflected light. Black tourmaline, white feldspar showing good cleavage, and quartz.

Consequently the sequence of changes in the granite seems to be:
1) disappearance of biotite;
2) deposition of tourmaline along grain boundaries, maybe principally at the cost of feldspar;
3) introduction of quartz, which replaces the feldspar and shows a remarkable preference for the heterogeneous tourmaline-feldspar contacts.

Photo 27, a twofold enlargement, under reflected light, of the polished surface of a tourmaliniferous pegmatitic rock (No 74, north flank Ihunga) shows clearly how the tourmaline is deposited along the grain boundaries of the feldspar crystals, or penetrates into these crystals particularly along cleavage planes. It is also evident that the quartz has been preferably deposited along the heterogeneous tourmaline boundaries replacing the feldspar.

The name pegmatitic rocks is used here exclusively in its actual significance, denoting relatively coarse-grained rocks. The opinion is gaining ground that the genesis of pegmatites in general is accompanied by the occurrence of solutions approaching the hydrothermal stage. Also the micro-structure of these rocks described above points in that direction, and, as far as can be made out with the aid of the magnifying glass, the same applies for the deposition of tourmaline, mica, cassiterite and the greater part of the quartz contained in the pegmatite dykes proper.

Away from the granite towards the contact with the sediments feldspar gradually disappears, and extreme silicification has resulted in the formation of true "greisens" almost entirely consisting of quartz and muscovite with a few grains of tourmaline and a few specks of feldspar. This rock will be reverted to in chapter 3.

CHAPTER 2

INJECTION SCHISTS

Along the edge of the granite bodies one finds, besides the pegmatitic rocks, also other remarkable rocks projecting wedge like into the granite (see photo 35 chapter 3, taken from the northern slope of the Ihunga). These rocks are found near the contact in the biotite granite and consist of layers of mica-quartz schists interlain with wedge-like tourmaliniferous veins of granitic composition. As regards the strike the foliation of this rock corresponds to the strike of the nearest sediments. The dip is very difficult to determine, but its direction is also equal to that of the nearest sediments. The layers of mica-schist are supposed to be remnants of phyllites. Microscopical examination seems to prove this assumption. The writer has designated them by the name "injection schists", in analogy with the term "injection gneisses", though the term infiltration schists might tentatively be suggested.

Photos 28 a and b taken from specimen No 119 n (north slope Ihunga, magn. 12x) represent a section cut perpendicular to the foliation; the part on the right is mica-quartz schist and that on the left granitic material, the latter being strongly impregnated with tourmaline. The sponge-like shape of the tourmaline is very characteristic and here again it is clear that tourmaline penetrates along the grain boundaries of the feldspar. Almost in the middle there is a fair example of this, and it is striking that each separate grain of microcline is entirely surrounded by tourmaline. These feldspar grains have subsequently also been protected, for *larger feldspar grains are only to be found in the tourmaline*, whilst *outside of it there are only extremely fine drops of microcline left in the quartz*. Where the tourmaline envelops quartz one mostly finds minute remnants of feldspar against the edge of the tourmaline. Here, too, a large quantity of quartz seems to have been added after tourmaline had already been deposited along the feldspar grain boundaries, and apparently both tourmaline and quartz have by preference replaced the microcline.

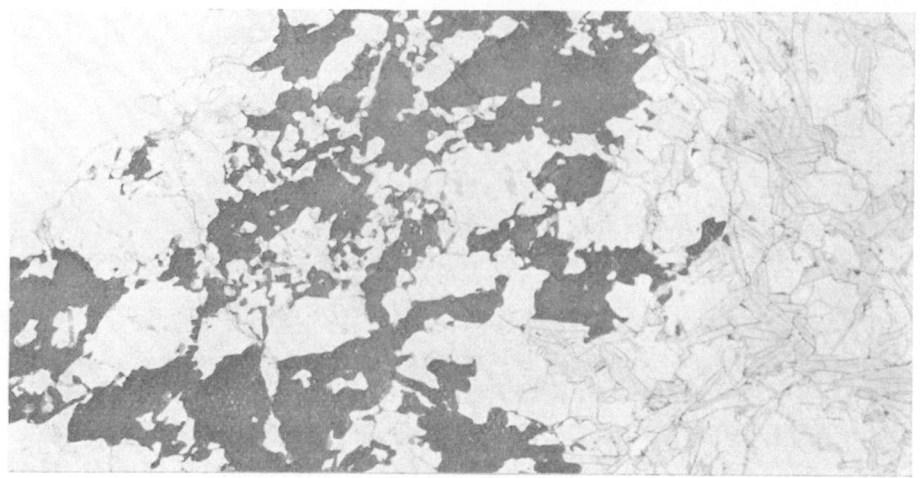

Photo 28a (No 119 N, magn. 12x, ordinary light). Shows tourmaline (black) embracing feldspar, quartz and mica. Remnant of schist to the right and schist replaced by granitic material to the left. (NE. spur of Ihunga).

Photo 28a also clearly shows that the heterogeneous grain boundaries in the quartz-mica schist are exclusively formed by the basal plane of the mica (001).

Photo 28b. Same as 3a under crossed nicols. Quartz embraces many small drops of microcline.

This applies not only to the mica-quartz but also to the mica-tourmaline contacts. Moreover the tourmaline often occurs along the homogeneous mica boundaries. It would appear, therefore, that mica already existed before tourmaline was deposited. Of tourmaline it is certain that this originated through the metasomatic action of magmatic emanations, although it is not known how much of the tourmaline substance is derived from the original rock, nor how much has been introduced from outside (except as regards Fluor and Borium).

For the matter of that this is immaterial, since the *tourmaline crystals* can only have been formed by metasomatism. Considering that their grain boundaries are always parallel to the predominating crystallographic planes of the mica it is logical to assume that they were formed subsequently to the mica. The same arguments raised in respect of quartz cannot be checked so definitely as in the case of tourmaline, for as yet there is no certainty regarding the amount and structure of the quartz originally present. In the recrystallization of a rock which yields quartz and mica without the substantial addition of new material one might, however, expect a sieve structure or a varying relation between quartz and mica, sometimes the quartz and sometimes the mica showing the characteristic features of the youngest mineral. This quartz-mica schist, however, is derived from the upper sericite phyllites, of which the silica content in the unchanged condition is about 48-50 %, and which almost entirely consist of sericite. There is no doubt that on recrystallization without the addition of fresh material these sericite phyllites are totally converted into mica. As a matter of fact in unchanged condition these phyllites mainly consist of sericite. Most probably, therefore, the greater part of the free quartz has been added subsequently, under simultaneous replacement of pre-existing mica.

Photo 29a (No. 113n, magn. 14x, ordinary light). Shows tourmaline (dark) containing inclusions of mica, the latter in some places replaced by quartz.

Photo 29b. Same as 29a under crossed nicols. One inclusion (near left-hand edge) in tourmaline contains quartz (nearly extinguished) and a small remnant of mica along its edge (bright). Some microcline is visible.

Photos 29a and b also give evidence of the quartz being younger than mica (No 113 n. N. slope of the Ihunga). In the tourmaline there is a small scale of muscovite, and near that an enclosed grain of quartz which is pseudomorphic after mica (near middle of left edge); further, as may clearly be seen, in this quartz there is a small remnant of mica against the top edge (quartz is practically extinguished and mica is bright). Apparently the quartz shows a preference for the muscovite rather than for the tourmaline and manages to reach the mica even where it is embraced by tourmaline.

In the conversion of microcline into quartz, which in these rocks took place subsequent to the deposition of tourmaline, large quantities of solutions carrying potassium and aluminium must necessarily have been liberated, so that it would not be surprising if also traces of a mica generation were found which appears to be younger than tourmaline. This is observed in the same slide No 113n (photo 30, magn. 14x); at the top there are numerous mica scales

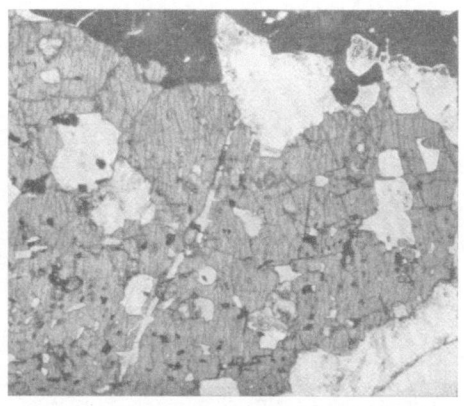

Photo 30 (No. 113n, magn. 14x, ordinary light). Shows tourmaline (dark or black) embedded in quartz. Roughly rectangular inclusions of mica and quartz visible in tourmaline.

enclosed in tourmaline, but these mica scales are bounded by a prismatic plane of tourmaline and the parting parallel to 001 of the latter. The muscovite is by no means intergrown parallel to the tourmaline. In the middle of the same photo a minute quartz veinlet cuts through the tourmaline, following the heterogeneous tourmaline-mica contacts. At the right-hand side also the quartz is seen to constitute rough rectangular inclusions in the tourmaline, the grain boundaries of which are likewise determined by the predominant crystallographic directions of the tourmaline. It cannot be said whether this quartz has replaced tourmaline itself or mica younger than tourmaline, but considering the observations mentioned above the latter possibility is not precluded.

Fig. 29 also gives evidence of a subsequent replacement of tourmaline by mica. The tourmaline, occurring in coarse crystals in autometamorphic granite, contains veinlets of mica which are bounded by crystallographic planes typical for tourmaline. Therefore, the replacement of tourmaline by mica is not restricted to injection schists alone.

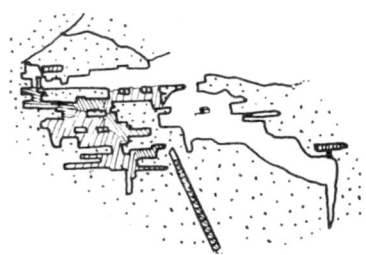

Fig. 29. Tourmaline (dotted), mica (hatched parallel to cleavage) and quartz (white).

Thus the metasomatic transformations in the rocks described in chapters 1 and 2 lead one to presume that soon after the consolidation of the granite, solutions containing F and Bo found their way upwards. Subsequent silicification resulted in the disappearance of feldspar and the resorption and corrosion of both tourmaline and mica. In consequence of the silicification of feldspar, solutions were formed which occasionally deposited a mica younger than tourmaline.

CHAPTER 3

THE FORMATION OF MAGMATIC SOLUTIONS

From the foregoing it is evident that replacement of feldspar by quartz is one of the most important autometamorphic processes of general occurrence. The silicification of feldspar results in the formation of solutions from which mica might be deposited. As a matter of fact these solutions have left evidence of their passage, as silicification in the granite is always accompanied by an increase of coarse mica. Though the deposition of quartz appears to be genetically connected with the deposition of coarse mica, observations in the Karagwe-Ankolian granite have brought to light the fact that most of the mica is invariably older than quartz.

These observations are proved by microscopical investigations, which show that quartz has been preferably deposited along homogeneous mica contacts. Moreover the quartz forms small veinlets in the mica parallel to the basal plane of the latter, and generally the heterogeneous quartz-mica contact is parallel to the cleavage plane of the mica. Often several booklets of mica, separated by quartz, are simultaneously extinguished under crossed nicols, and

Photo 31 (No U 74, magn. 12x, under crossed nicols). Silicification of plagioclase by quartz.

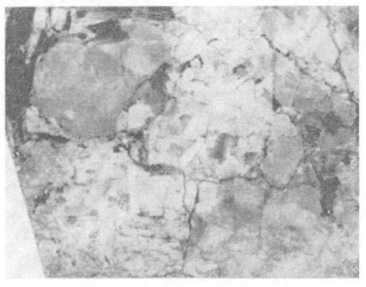

Photo 32 (No U 74, magn. 2x, reflected light). Silicified granite forming nothern wall rock of pegmatite dyke (cf. map 4). Feldspar (white), quartz and coarse plates of mica are visible.

where two booklets abut at an arbitrary angle quartz forms small veinlets along the homogeneous contact.

These features are clearly visible in the "greisens", which are particularly situated along the contact with the sediments. The transition from "greisen" into granite is formed by an aureole of pegmatitic rocks, described in chapter 1.

Consequently, the plane of contact appears to have provided an easy pathway for the ascending solutions, these having deposited large quantities of quartz. This latter mineral was preceded by abundant coarse mica, which was partly resorbed during the silicification.

In the "greisens" remnants of tourmaline and feldspar may occasionally be observed. The latter mineral occurs in diamond-shaped specks, the outline of which obviously corresponds to the cleavage planes of the feldspar. These specks are grouped together and often appear to constitute the remnants of one larger crystal, for the cleavage planes of several specks are parallel, as may also be seen from photo 31.

Photo 32 gives the twofold magnification of the polished surface of a silicified granite (reflected light).

Photo 33 (No 103, magn. 12x, crossed nicols). Silicification in Rubanda granite. Microcline, plagioclase and quartz are visible.

In the Rubanda granite the silicification gave rise to the formation of peculiar intergrowths of quartz and feldspar, which at first sight might be taken for eutectic intergrowths (photos 33 and 34). In photo 33 the core of plagioclase lath is partly replaced by quartz. In the microcline, however, small tubular veinlets have penetrated along the cleavage planes of the feldspar. The number and dimensions of these veinlets noticeably increase towards the border of the microcline crystal (photo 34).

Photo 34 (No 103, magn. 12x, crossed nicols). Rubanda granite showing pseudo-eutectic intergrowth between quartz (bright) and microcline.

The contact between quartz and feldspar is often parallel to the direction of one of the cleavage planes of mica, which fact is illustrated in fig. 30.

This silicification is also visible in the injection schists. Photo 35 shows the contact between a schist layer and a microcline crystal. Many mica folia are enveloped by a film of quartz which has been deposited along the heterogeneous boundaries.

A small veinlet of quartz cuts through the left-hand corner of the microcline, following the twinning planes.

Another example of silicification is supplied by the only aplite found in Wishikatwa.

This dyke is situated a little to the west of the Kavusanammi (hill

Fig. 30 (No 102, Rubanda). Microcline (cross-hatched parallel to cleavage) with veinlets of quartz (white).

Photo 35 (U 111, magn. 14x, crossed nicols). Shows contact between microcline and remnant of recrystallized phyllite. Films of quartz (bright) envelop mica.

No 30 Wishikatwa), between the steeply overtilted sediments and the silicified granite (cf. photo 32, part IV A).

This ap'ite dyke is only 1 to 1½ metres wide and is also short, as it is no longer to be seen in the two sides of the ravine. It lies in a sort of cul-de-sac, where the granite penetrates deep into the sediments; it forms a steep wall which in the rainy season causes a small waterfall.

The dyke consists of a very fine-grained rock of a yellow colour studded with red euhedral garnets. Tubular quartz veinlets with an angular section are found in the ground mass. Under the microscope the fine-grained ground mass appears to consist of idiomorphic to hypidio-morphic laths of plagioclase, the composition of which lies between that of oligoclase and albite; photo 36 is a fair representation of this rock (No U. 77 magn. 18 x). This plagioclase in some places forms extremely fine-grained aggregates. The sections may be seen of three tubular quartz veinlets, and again it is clear that their cross-sections are for the greater part bounded by M of the plagioclase. In the quartz in the top right-hand corner of the photo the same sort of small rounded inclusions of feldspar may be seen as have been mentioned in chapter 1, but here they show albite lamellae. This,

Photo 36 (U 77, magn. 18x, crossed nicols). Aplite dyke west of Kavusanammi-West (cf. map 4). Garnet in middle of right-hand edge. Further plagioclase with three veinlets of quartz.

together with the tubular shape of these veinlets and their boundaries, makes it seem quite probable that the quartz has penetrated by replacement of feldspar.

The formation of the coarse crystal of basic plagioclase, which is seen in the above mentioned photo in a state of extinction to the left of the garnet, may also be ascribed to metasomatism.

The mica-depositing solutions have left their traces not only in the sediments but also in the granite. The existence of mica younger than feldspar may be noticed in the granite.

Photo 37, for instance, shows clearly the intergrowth between mica and the plagioclase. The heterogeneous boundaries are parallel to the cleavage planes of the feldspar, but never parallel to the basal plane of the mica. The conclusion to be drawn from this is that the feldspar was present at the time the mica was metasomatically deposited.

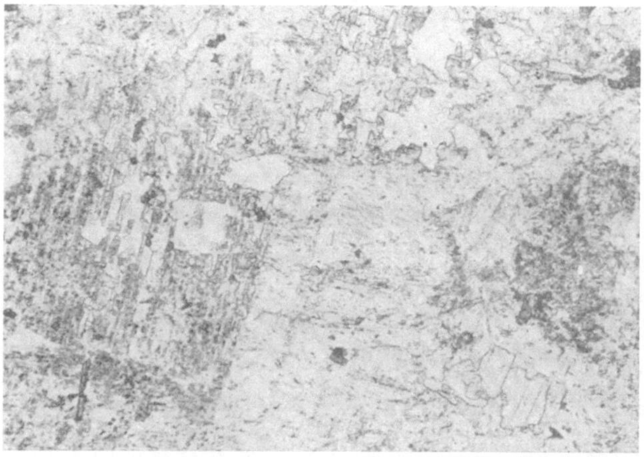

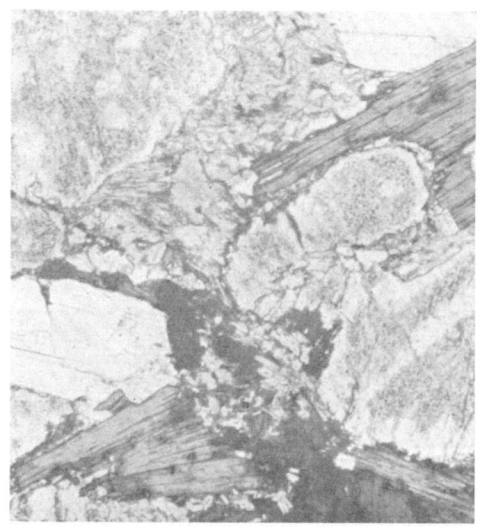

Photo 37 (No 81, magn. 31x, ordinary light). Intergrowth between mica and plagioclase.

In photo 38, which was taken from a slide of the Rubanda granite, it is to be seen that the biotite is more or less bent and partly altered into chlorite. The muscovite, however, lacks evidence of such bending. The heterogeneous mica-biotite boundary is formed by the basal plane of the latter. The white mica, therefore, proves to be a younger constituent of the granite and it may have been deposited at the cost of feldspar (plagioclase).

As silicification has affected both plagioclase and albite as well as microcline, one may presume that the solutions which were subsequently formed carried potassium as well as calcium and sodium. Potassium, however, will have been the predominant constituent, as microcline prevails in the granites over the other members of the feldspar group.

Moreover, there is evidence that apart from silicification potassium is liberated, sodium taking its place. Photo 39, for instance, shows the presence of albite in microcline, the albite forming parallel veinlets in the potassium feldspar. This example of albitisation has been observed in specimens from the Katerero pegmatite dyke. In this rock a subsequent silicification is clearly visible and may also be noticed in the above-mentioned photo, where the cross-section of a tubular veinlet

Photo 38 (No 103, magn. 31x, ordinary light). Rubanda granite, showing partly altered biotite (dark) with numerous inclusions of zircone. Further muscovite and plagioclase (turbid).

of quartz is present. These veinlets are abundant in the very coarse lumps of microcline which constitute the dyke. Where silicification is very intense coarse plates of white mica occur, as for instance near the Chobugombe river (cf. map 4).

Another origin of potassium-bearing solutions may be sought in the extensive formation of myrmekite, occurring in all varieties of granite and auto-metamorphic rocks derived from granite.

Excellent examples of this have been shown in photos 19 and 21, chapter 6, part II. In these cases Ca and Na have been substituted for potassium, while for each Ca molecule a corresponding quantity of quartz has been liberated.

It must be emphasized that the myrmekite shown in the above mentioned photos has not been formed along the heterogeneous contacts between plagioclase and microcline. In the Ankolian granites even the greater part of the myrmekite occurs as totally independant formations in the microcline. Generally it is found along the edges of the microcline and its grains are often bounded by the cleavage planes of the host mineral.

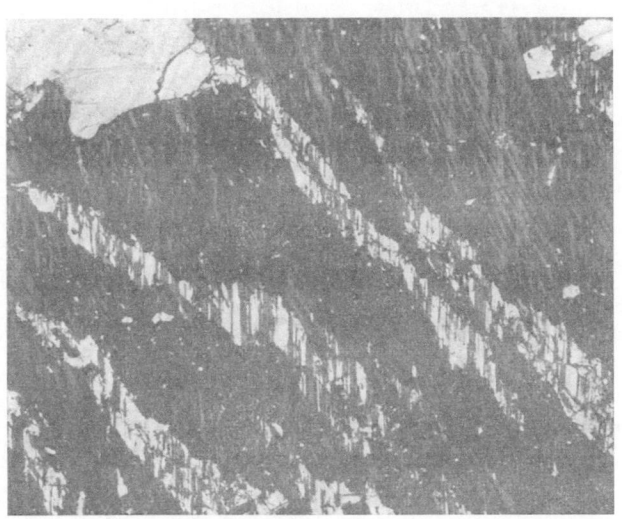

Photo 39 (U 41, magn. 12x, crossed nicols). From Katarero pegmatite dyke, showing veinlets of albite in microcline. Cross-section of a tubular veinlet of quartz in top right-hand corner.

Photo 40 shows further examples of the formation of myrmekite invariably occurring in microcline; numerous small grains of the latter are almost entirely replaced by myrmekite.

The formation of myrmekite is undoubtedly a special form of albitisation. It is the result of metasomatic action of mainly sodium-bearing solutions on potassium feldspar. These solutions

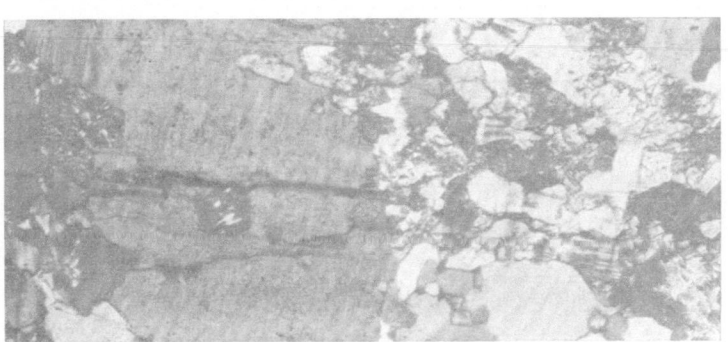

Photo 40 (U 64, magn. 31x, crossed nicols). Fine-grained marginal variety of biotite granite, showing grains of myrmekite in microcline; quartz also present.

contained a certain quantity of calcium, which element might have partly originated from the replacement of plagioclase by white mica (cf. photos 37, 38 and 41). Indeed, where myrmekite is abundant, numerous younger mica scales are found in the old laths of plagioclase, whilst microcline very rarely shows evidence of such replacement. In Ankolian granites also rims of myrmekite around plagioclase laths are observed, where the latter is embedded in microcline (photo 41). These laths contain many scales of white mica and consist of a more calciferous plagioclase than the myrmekite rim, a common fact which has been mentioned by many authors. Therefore it might seem as if part of the potassium of the microcline, being replaced by sodium, had been deposited as muscovite[1]) in the plagioclase, whilst the calcium from the latter was incorporated by the albite rim under simultaneous depositon of free quartz. Anyhow microcline disappears, the sodium content of the rock increases and white mica becomes more abundant.

Photo 41 (No 20b, magn. 31x, crossed nicols). Shows rim of myrmekite around plagioclase embedded in microcline. Plagioclase core contains numerous scales of biotite and is partly replaced by younger (white) mica.

The final conclusion may be formulated as follows:

Autometamorphic changes in the granite mostly occur along the plane of contact with the sediments. F and Bo-bearing solutions or gases have deposited tourmaline, which mineral has mainly replaced feldspar.

A subsequent deposition of white mica is in turn followed by an intense silicification and the deposition of coarse mica appears to be in genetic relation to the latter process.

Deuteric processes are of more general occurrence and mainly result in the disappearance of potassium feldspar, which is converted into albite or myrmekite, whilst potassium has been carried away.

A certain part of this potassium may have been deposited in the granite as muscovite, especially at the expense of plagioclase.

From the evolution of the magmatic solutions in the granite one might expect the following sequence of deposition in the sediments: first tourmaline, then mica and finally quartz. The deposition of tourmaline, however, may have been preceded by other processes.

---

1) However, it is still doubtful whether the white mica in plagioclase consists of muscovite or of paragonite.

B. ALLOMETAMORPHISM CAUSED BY MAGMATIC SOLUTIONS

CHAPTER 4

INTRODUCTION — SEQUENCE OF DEPOSITION

In the foregoing pages the author merely wished to trace the origin of some of the most important radicals dissolved in the magmatic solutions. Also TiO , SnO must have been present, seeing that rutile [1]) and cassiterite are found in the contact metamorphic aureole of the granite [2]). Afterwards the nature of the solutions changed and mispickel and hematite were deposited.

The nature of the solvent and the presence of other dissolved substances will not be discussed here, as they have left no traces in the rocks now to be described.

It is the aim of the author to follow in this part the traces left by the magmatic solutions in the Karagwe-Ankolian rocks on their way upwards.

Joints, faults or disturbed zones, bedding planes or even subcapillary rents will have served as a pathway, and as a matter of fact the evidence of their passage can be observed most easily either along fault planes or quite near the contact with the granite. Where open fissures were not available the solutions will more or less have filtered through the whole rock and especially bedding planes and grain boundaries of the rock minerals will have provided the pathways. Osmotic pressures may have been caused by differences in concentration in different parts of the invaded rock which were in communication with each other, and also these may have had some considerable influence on the course of the solutions.

The distribution of additive metamorphism caused by the ascending magmatic solutions will depend on the course of their channels and consequently the metasomatic minerals deposited by these solutions will be found in local accumulations.

This fact may serve as a criterion to establish the hypogene origin of a metasomatic mineral, for when without addition of new materials and solely in consequence of altered conditions (temperature, pressure, stress, etc.) the

---

[1]) often developed as geniculated twins.

[2]) Wolframite has been found in stanniferous deposits near Kamwezi, outside the area explored by the author. Apatite is frequently observed in the vein quartz.

original components of a rock are re-grouped in a newly formed mineral, this metasome will be regularly distributed throughout the rock.

Thus the garnets of the garnet phyllites, which are regularly distributed in this belt, are certainly not of hypogene origin, while irregular occurrences of tourmaline and mica and local accumulations of quartz must be attributed to additive metamorphism. The more so as the Karagwe-Ankolian sediments have a very uniform composition which remains the same in spite of great distances.

Metasomatic changes take place on the basis of equal volumes. Molecular volumes, therefore, will play an important part and simple chemical equations cannot hold good.

Nevertheless the formation of a metasomatic mineral appears to be governed by definite, though intricate, chemico-physical laws, and conditions of temperature and pressure, the nature of the solution and the chemical and mineralogical composition of the country rock have played an important role.

From field observations and microscopical investigations these rules may be empirically deduced, for only such rules can account for the fact that in Ankole for instance tourmaline has mainly been deposited in the close vicinity of the granite, whilst, other conditions being equal, it is more abundant in aluminous rocks (e.g. sericite phyllites) than in more sandy rocks (e.g. Middle Division), and that along a former main channel of the solutions its greatest abundance is found over a limited vertical extent.

As regards the distribution and shape of metasomatic minerals, the microscopical investigation of Karagwe-Ankolian specimens has invariably shown the two following rules:

1. The metasome has been preferably deposited along the grain boundaries of the minerals that were present at the time of its deposition.
2. The development of the metasome has been influenced by the crystal structure of the palasome.

It must be noted, however, that only silicates were considered.

Further it is not to be said that a metasome cannot develop in euhedral grains. As far as the author's investigations go, it has been observed that for a given mineral the degree of idiomorphism is greatly dependant on the intensity of metasomatism (apart from other factors, such as the nature of the palasome).

But it could generally be stated that even idiomorphic minerals show somewhere along their edges the influence of the important crystallographic planes of the replaced mineral. Since this is of great importance for determining the sequence of crystallization it will be gone into further. In order to demonstrate the problem by a concrete case the reader is referred to fig. 31 (No 18, Kaina), representing an intergrowth of cassiterite and mica. Is it to be assumed here that the cassiterite replaced the mica, or that the mica replaced the cassiterite?

Fig. 31 shows quite clearly how the cassiterite penetrates [1]) into the mica aggregate between three coarse mica folia, being bounded for the greater part by 001 of the mica. On examining the mica folia extending into the cassiterite it is evident that neither their position nor their outline has any causal connection with the important crystallographic planes of the cassiterite; there is no mica on the cleavage and twinning planes of the cassiterite, as would be expected if the cassiterite had been replaced by mica. A very striking difference is to be noted in fig. 32 (No 14b, Kaina), in which on the left-hand side the cassiterite grain is clearly bounded by 001 of the mica, whilst the cassiterite projects

Fig. 31. No 18 Kaina.

into the mica along the homogeneous grain boundaries of the latter. This figure shows twinning and cleavage planes of the cassiterite along which there is no mica. This is not the case with the quartz, which forms a veinlet, the outline of which is determined by the twinning plane and the cleavage of the cassiterite. Consequently, until observations made elsewhere prove the contrary, it is reasonable to assume that the

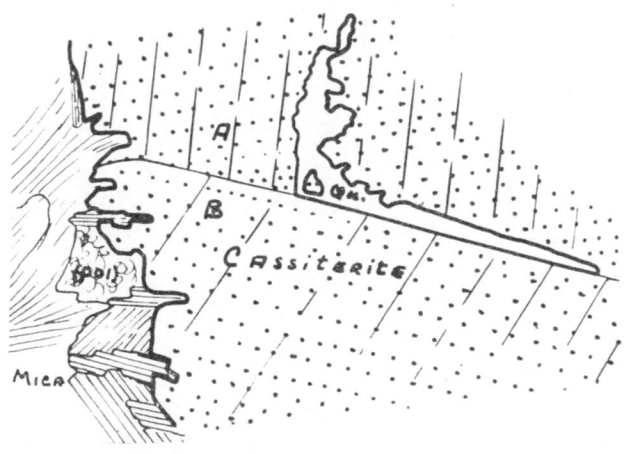

Fig. 32 (No 14 Kaina). Shows cassiterite twin intergrown with mica and partly replaced by veinlet of quartz.

sequence of deposition was: mica — cassiterite — quartz. It may be argued that the foregoing only proves that mica forms euhedral crystals and quartz xenomorphic crystals, so that cassiterite may well be older than quartz and mica. A glance at fig. 33 (No 38b, Kashozo), will show that though tourmaline occurs in the form of fairly euhedral to hypidiomorphic crystals, many of which are intergrown parallel to the mica, its shape and distribution are determined by the edges and cleavage planes of the mica. From this the author concludes that tourmaline has replaced mica. Turning now to the cassiterite (fig. 33), the periphery of this is bounded for the greater part by the basal cleavage plane of the mica, and in the

Fig. 33 (No 38 Kashozo) = Diameter about 1 mm. Intergrowth of mica (hatched parallel to cleavage), tourmaline (dotted) and cassiterite with cleavage and cracks (c). Top right-hand edge quartz (white).

---

1) This term is merely used for the sake of easy description.

Stheeman, Geology.

middle of the figure the cassiterite is seen projecting [1]) in a veinlet between two mica scales, both sides of the veinlet being bounded by 001 of the adjoining mica. After reaching the tourmaline grain this veinlet bends almost 180 degrees and continues between the mica and the tourmaline. Further it is to be emphasized that the heterogeneous tourmaline-cassiterite boundary is parallel to the cleavage plane of the mica scale lying on the other side of the cassiterite. Consequently cassiterite is in any case younger than mica and tourmaline, so that the author takes the probable sequence to be: mica — tourmaline — cassiterite. At the bottom of the figure it is also clearly seen how the cassiterite penetrates along the tourmaline grain.

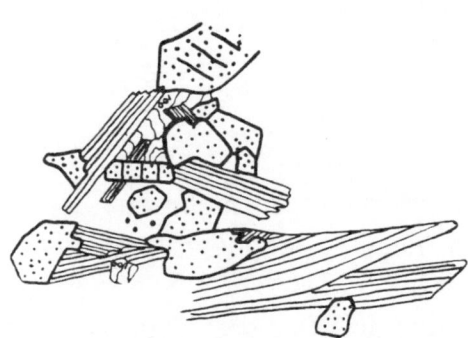

Fig. 34 (No 38 Kashozo) = Diameter about 1 mm. Mica (hatched parallel to cleavage) and tourmaline (dotted, with basal parting) embedded in quartz.

Finally the reader is referred to fig. 34 (No 38a, Kashozo). The tourmaline grain in the bottom left-hand corner shows an entirely idiomorphic boundary on the left side, whilst on the right side a small protuberance projects in between two mica scales. All the other tourmaline grains likewise appear to be bounded to a large extent by the course of the homogeneous mica grain boundaries and the direction of the mica cleavage. The conclusion to be drawn from this is that in the Karagwe-Ankolian specimens no mineral appears to have possessed such a crystallization power as to disregard entirely the crystal structure of the palasome. The same will be observed over and over again and it can therefore safely be assumed that the sequence of deposition of metasomatic minerals can be established in the manner demonstrated above.

On the whole mica was found to be no exception, for it has already been shown in the discussion of the granites that mica may also be bounded by the cleavage planes of the palasome (photos 30 and 37 part IV A).

CHAPTER 5

SERICITE PHYLLITES IN THE VICINITY OF GRANITE

As stanniferous veins are invariably associated with the formation of tourmaliniferous phyllites or schists, only these rocks will be discussed here. Moreover the omission of other types of contact metamorphic rocks is fully justified, as they do not show any remarkable features.

---

1) see footnote page 81.

Tourmaline is found in the sediments in the immediate vicinity of the granite. It is also observed at a greater distance from the granite but in that case it is deposited along fault planes or fractures, and the presence of quartz-mica veins along these faults gives another indication for the passage of considerable quantities of solutions.

Tourmaline is abundantly found in the sericite phyllites, undoubtedly because this horizon is always situated close to the granite, but also because its chemical and mineralogical composition were favourable for the deposition of this mineral. Indeed it is seldom found in the garnet phyllites, in the Ihunga quartzite or in the arenaceous phyllites and sandstones of the Middle Division, a phenomenon which is not due only to the fact that these belts generally lie at a great distance from the granite.

This can be seen where these belts abut against the granite, as for instance on the Ruitonbero.

In the garnet phyllites exposed on this hill the predominant contact-metamorphic minerals are staurolite and mica, while in the sericite phyllites at the foot of this hill tourmaline prevails. In the Ihunga quartzite, to the north of the Ruitonbero, no metamorphic minerals can be detected.

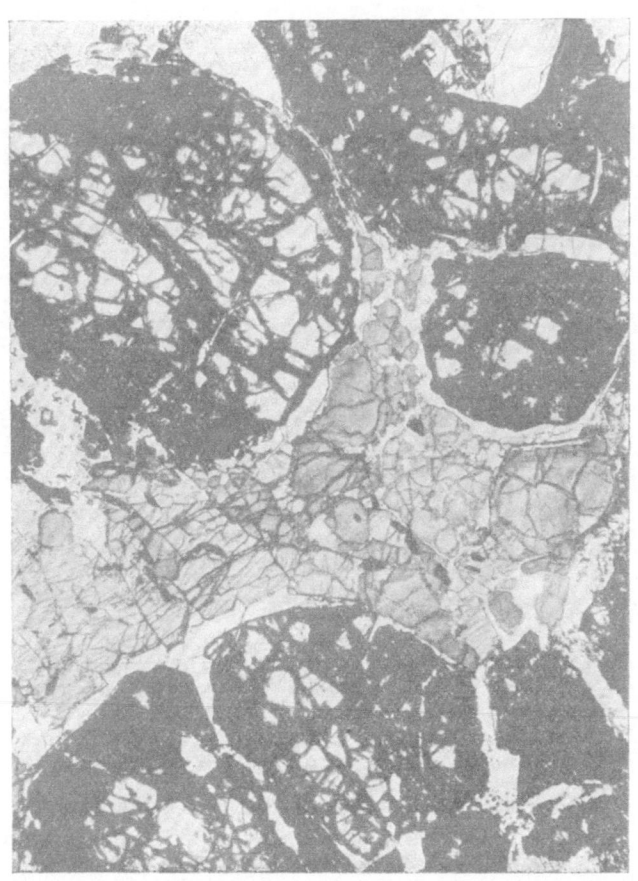

Photo 42 (No U 31, magn. 16x, ordinary light). Tourmaliniferous garnet phyllite on hill A., Katuba valley; weathered garnets partly replaced by limonite; tourmaline and interstitial quartz.

Where the mixed clayey-arenaceous sediments of the Middle Division form the contact with the granite (e.g. in the Kavungo syncline) generally staurolite and mica are the only contact metamorphic minerals present, and investigations of A. J. Speyer have clearly shown that tourmaline is scantily present in that area. In this area the author observed tourmaline only near the stanniferous vein at the Chamgash and near the Kashozo vein, whilst Speyer indicates on his map only a few more occurrences.

In the garnet quartzite it has never been found, nor in the equivalent arenaceous belt, except on Buramma Ridge and Ruechimarra hill, where its formation is undoubtedly due to the existence of a reverse fault.

The occurrence of tourmaline between hills 36 and 37 (Katuba valley) in the garnet phyllites far from the granite (cf. photo 42) is also due to the existence of faults, which have been proved on that particular spot; some of them are marked by the presence of quartz lenses.

Another occurrence of tourmaline in these arenaceous belts is found in the Ihunga quartzite in the axis of the Katuba syncline (between hills No 30 and No 31). Probably the granite has intruded in the axis of that syncline and the vertical distance from the granite may be small. The fact, however, that tourmaline is very abundant here must also be sought in the tectonics of this syncline, which has suffered several phases of folding. The same explanation must be applied to the occurrence of tourmaline in the garnet phyllites on the southern spurs of the Kavusanammi (cf. map 5).

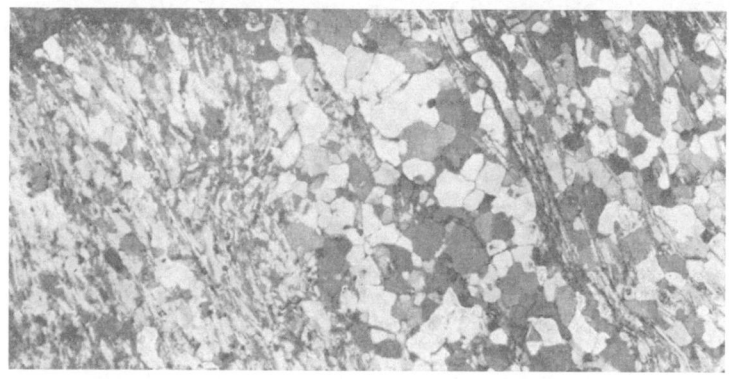

Photo 43 (No 83, magn. 17x, under crossed nicols). Shows irregular distribution of quartz in sericite phyllites in a specimen of lower sericite phyllites at Kabezi. Left-hand edge little quartz — middle small quartz lens — right-hand side few scales of mica.

Therefore it appears that not only the vicinity of the granite or the composition of the mother rock governs the degree and type of additive metamorphism, but also faults may be the cause of mineralization being found at a great distance from the granite or in normally poorly mineralized belts.

The sericite phyllites generally have been intensively altered. These bluish-grey silky phyllites, which in unaltered condition are almost entirely composed of sericite, have been more or less completely transformed into tourmaliniferous mica schists. Coarse mica is abundantly observed along the bedding planes or in irregular lenses. Everywhere near the granite silicification is apparent. The formation of irregular lenticular intercalations mainly consisting of quartz intergrown with varying quantities of mica (photo 43) must be due to additive metamorphism, as the belt of sericite phyllites is uniformly poor in free silica when in the unaltered condition.

On the whole it may be noted that the degree of metasomatism is extremely variable, and strongly tourmaliniferous or silicified strata are interlain with or merge into slightly altered phyllites.

Three specimens of the tourmaliniferous metamorphic sericite phyllites will be described here.

One specimen is represented in photo 45 (No U. 109, upper sericite phyllites

west of the Ampazzo; it also shows a quartz-mica veinlet, which will be referred to in part V); another specimen is shown in figs. 35—40 (specimen No 100,
lower sericite phyllites, Buramma Ridge, close to points A, and C). Sericite
phyllite in slightly altered con-
dition is shown in photo 44. A
few prisms of tourmaline are
present, while the major part of
the slide consists of plicated
sericite. Silicification is obvious,
for here and there lenses and
grains of quartz increase in number to such an extent that only a
few remnants of mica are left. The
striking inhomogeneity of the
rock denotes additive metamorphism.

With the aid of a few sketches
a clear idea can be obtained of
the sequence of deposition of the
various constituent minerals.

Photo 44 (No 72, magn. 31x, ordinary light) Slightly
tourmaliniferous sericite phyllites of Kaina. Shows mostly
plicated sericite with a few crystals of tourmaline and scales
of hematite. Top edge partly silicified. On the bend of the
minute folds small grains of quartz.

Figs. 35—38 (all from specimen
No 100b) show the heterogeneous
grain boundary between tourmaline and mica. It is seen how in some places the
tourmaline is intergrown parallel to mica, the former being present as hypidiomorphic prisms. The grains of tourmaline are, however, invariably bounded
particularly by the basal plane of the mica, and where this is not so there may

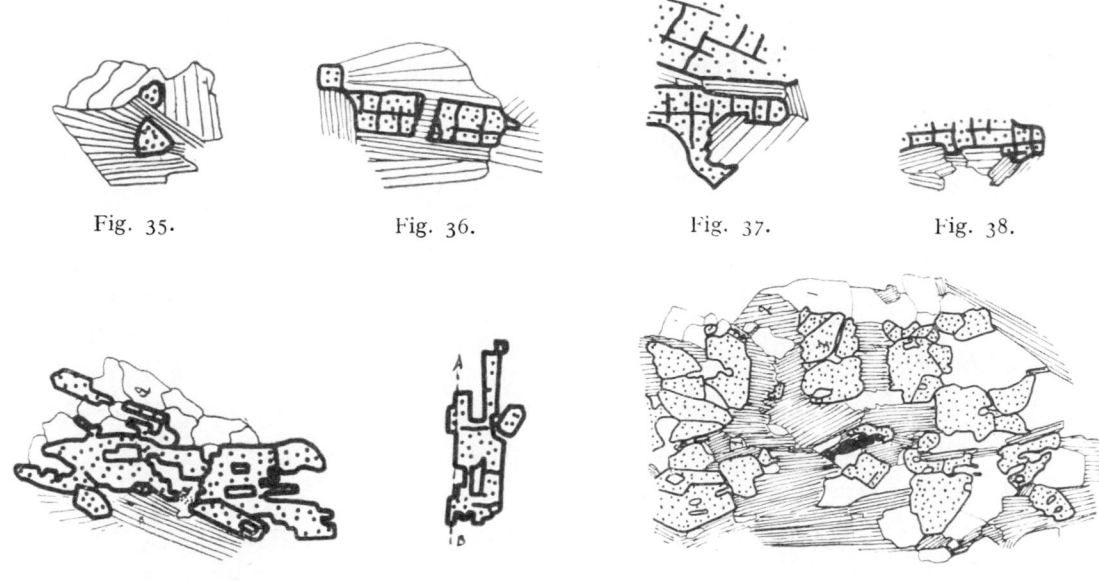

Fig. 35.   Fig. 36.   Fig. 37.   Fig. 38.

Fig. 39.   Fig. 40.   Fig. 41.

Fig. 35 to 41   Mica (hatched parallel to cleavage), tourmaline (dotted), quartz (white).

be seen a sawlike line of contact. The assumption that this does indeed arise from the replacement of mica by tourmaline and not by the reverse action — whereby mica would then tend to develop euhedral crystals — is supported by the fact of the *tourmaline grains lying practically everywhere on the mica boundaries, whilst hardly ever does a tourmaline grain occur in the middle of a mica plate.* Tourmaline indeed is particularly found along the edges of the mica plates. Further the outline of the tourmaline crystals is for a large part bounded by the arbitrary shape of the mica booklets (see figs 37 and 39). All these facts can only reasonably be accounted for when tourmaline has replaced the mica.

The tourmaline-quartz contacts, however, give quite a different aspect (see figs 39 and 40). The outline of the tourmaline is in nearly every case determined by its prismatic zone

Photo 45 (No U 109, magn. 13x, crossed nicols). Tourmaliniferous mica schists with small veinlet consisting of quartz and mica, the latter slightly coarser in the veinlet than in the schist.

Photo 46 (twofold magn., polished surface, reflected light). Aggregate of coarse mica metasomatically developed in upper sericite phyllites (South flank of hill 42 Katuba valley).

or by its parting according to 001. A typical feature is the crenelated form of the tourmaline. Quartz is often enveloped by tourmaline, but the periphery and position of these inclusions are not fantastic or arbitrary, the longitudinal direction of the inclusions being parallel to the prismatic zone. The shape of these inclusions is rectangular, the outline being parallel to a prismatic face and the parting plane of tourmaline.

In view of the peculiar outlines of the tourmaline it is not surprising that quartz should be found amid tourmaline; if the slide had been placed for instance through line A-B in fig. 40 the same picture would have been given as that shown in the actual slide. The relation of quartz to mica is quite clear in fig. 41; quartz also shows a preference for the mica boundaries, whilst the boundary of the quartz grains is for the greater part parallel to the basal plane of the mica.

The result is, therefore, that all the mutual relations point to the following sequence: 1) mica, 2) tourmaline, 3) quartz. It has to be added that mica may have originated from the recrystallization of sericite and that possibly a very small part of the quartz may have been present as a primary constituent of the sedimentary rock; this part, however, is so small as to escape notice. The microscopic picture in photo 45 agrees exactly with what has been observed from the Buramma hand specimens.

These rocks have a lepidoblastic texture (see photo 45) and are mostly highly lustrous and of a light colour, whilst the tourmaline occurs in the form of well-shaped needles or round grains lying parallel to the (imperfect) foliation. Small and coarse rosettes and books of mica are of frequent occurrence in the sericite phyllites (cf. photo 46). Rarely is the rock of a darker colour, owing to an excess of tourmaline.

CHAPTER 6

METAMORPHISM ALONG FAULTS

FORMATION OF METASOMATIC VEINS

It has already been emphasized that fault planes play an important role in the distribution of metamorphic rocks and of metasomatic minerals.

A clear example of this has been mentioned in a previous chapter (tourmaline north of hill "a" Katuba valley).

Another good illustration is provided by Kaina Hill. The occurrence of metamorphic phyllites in this particular spot cannot be ascribed to contact metamorphism, as faulting only brought consolidated granite and unaltered sericite phyllites into contact with each other (cf. part III, fig 27). The stanniferous veins, and the accompanying metamorphic aureoles bear a genetic relation to the fault.

The additive metamorphism, therefore, is absolutely due to the presence of faults. As these faults offered an easy pathway for the ascending magmatic solutions, the metasomatic processes are confined to the immediate vicinity of the fault. This is illustrated by the fact that at some distance from the fault not a trace of metasomatism can be detected in the unaltered phyllite.

Towards the fault plane the degree of metamorphism rapidly increases.

Some metasomatic minerals occur, of course, at a much greater distance from the main pathway of the solutions than others and among these tourmaline and quartz must be mentioned (photo 44), whilst the latter mineral undoubtedly occupies the most extensive area of deposition (photo 47). The sericite phyllites, which are the least affected by additive metamorphism, consist of sericite, while quartz is very irregularly distributed and locally replaces part of the phyllite (photo 47). Hematite is present in numerous scales, possibly a result of hydrothermal impregnation.

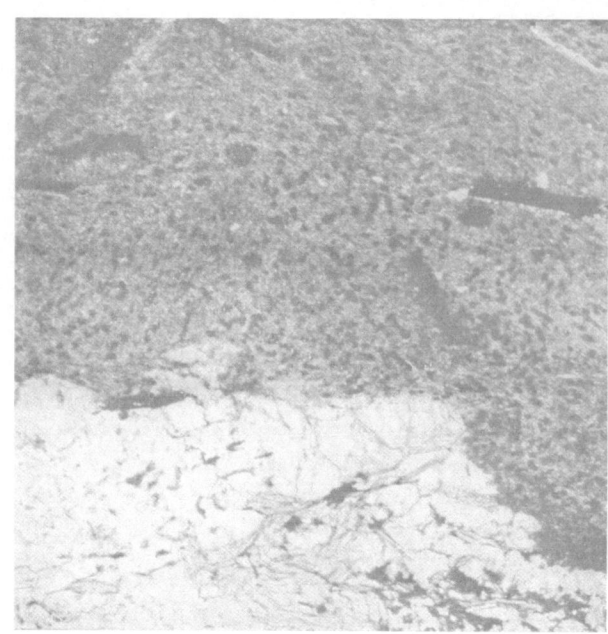

Photo 47 (No 23 Kaina, magn. 31x, ordinary light). Sericite phyllite with numerous scales of hematite. Bottom part almost entirely replaced by quartz, only a few scales of mica having been left.

When approaching the veins or fault planes these phyllites merge into a brownish-yellow, tourmaliniferous, lepidoblastic rock consisting mainly of mica. In the immediate vicinity of the lenses this rock changes into dark-green schists consisting for a large part of tourmaline. The texture of these very fine-grained rocks is nematoblastic, and in the arrangement of the very fine tourmaline prisms the same minute plication is sometimes to be seen as is noticed in the unaltered sericite phyllites. Also a reticulated arrangement of the tourmaline needles may be found (cf photo 48)

Quartz is, on the whole, less abundant than in the metamorphic rocks described in the previous chapter. It is also less irregularly distributed. Silicification is particularly very intense along the fault planes and in the veins, though the latter are only partly silicified. One of the lenticular veins (lens A), for instance, shows strong evidence of silicification

Fig. 42 (No 26 Kaina). Mosaic of tourmaline (white, only grain boundaries having been sketched) with interstitial mica (hatched parallel to cleavage or shaded when cut parallel to basal plane.)

along the roof, where also the metamorphic wall rock has been partly replaced by quartz; its floor, however, has only been slightly affected by silicification, and there quartz is a minor constituent.

The wall rock of the stanniferous lenses is a green schist. It consists for the major part of a strongly pleochroic tourmaline, the colour of which in transmitted light varies between brownish-white and dark greenish-brown, or in a few cases brownish-green. *This variety is always found in metamorphic sedimentary rocks as contrasted to granite.* Sometimes, however, a zonary arrangement of colour is observed, that in the core of the prism being bluish-green.

Fig. 43 (No 16) Mica (hatched parallel to cleavage), much coarser than in fig. 42, intergrown with tourmaline (dotted).

The interstices are filled with a very finescaly mica or rather sericite. The behaviour of the tourmaline towards the mica is sketched in fig. 42. Locally, however, the interstices are filled with quartz. As the vein is approached the mica becomes coarser. Fig. 43 (specimen No 16) shows quite clearly that

Fig. 44. Same as fig. 43.

also here tourmaline still occurs on the homogeneous mica boundaries and that its outline is often parallel to the basal cleavage of the mica.

Close to the vein, however, this changes — particularly where the mica becomes much coarser — and, as shown in fig. 44 (No 63b), it is seen that the tourmaline boundary is marked by its prismal zone and its parting parellel to 001, whilst the mica abuts arbitrarily against

Photo 48 (No 16, magn. 17x, ordinary light). Contact between vein and wall rock at the floor of lens A, Kaina. Tourmaline in wall rock but not in coarse mica of vein.

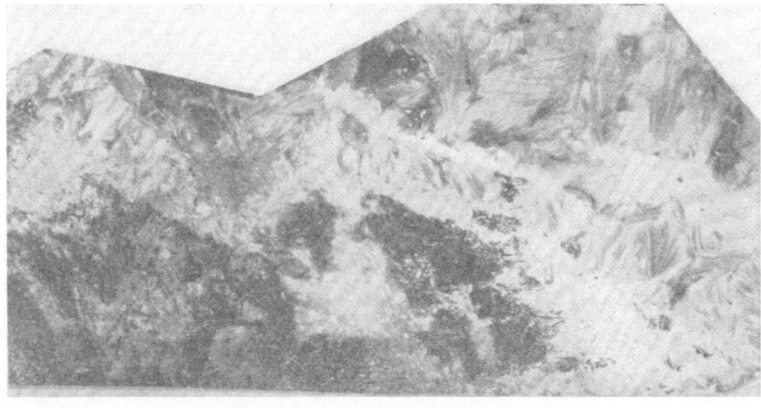

Photo 49 (magn. 2x, polished surface, reflected light). Contact at floor of lens A, Kaina. Tourmaline abundant in wall rock but lacking in coarse aggregate of mica constituting the vein. Some tourmaline prisms only present in medium-scaly mica. Cassiterite (black) in coarse mica in middle of top edge.

the tourmaline.

With increasing dimensions of the mica folia, the tourmaline becomes less and less abundant and in fact in the vein, which is composed of very coarse mica, no tourmaline at all is present (cf. photos 48 and 49).

Photo 49 gives a twofold magnification under reflected

light of the floor contact of the Kaina vein; it is clear that there is no tourmaline in the coarse mica, whilst it occurs only sporadically in the medium-

scaly mica. Finally photo 50 shows how small wedges consisting of mica and devoid of tourmaline occur in the wall rock.

Consequently the sericite and fine-scaly mica appear to be older than tourmaline. The deposition of mica, however, continued and the very coarse folia which were

Photo 50   (magn. 2x, reflected light, polished surface). Irregular contact of lens B, Kaina.

finally deposited particularly near the main channel of the solutions prove to be younger than tourmaline. During the last stage of the formation of mica, tourmaline has been resorbed.

As already stated, quartz is contained in the green tourmaline schists, and in the wall rock (at the floor of lens A for

Fig. 46.

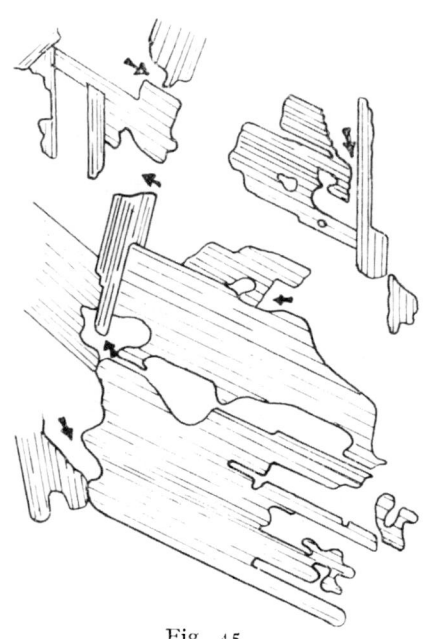

Fig. 45.   Fig. 47.

Figs. 45 to 47 (No 64, Kaina). Mica (hatched parallel to cleavage) and quartz (white). Latter particularly occurring along homogeneous mica boundaries (at arrow-heads).

instance) this quartz occurs in the interstices as a hard cement. So far the shape of the tourmaline is little affected by the presence of the quartz. It is particulary the mica that is corroded, and aggregates of scales bounded exclusively by ooi lie embedded in large quartz individuals (cf. fig. 46). Often one quartz grain embraces poikilitically a number of fine mica scales.

It might be supposed that fine-scaly mica has replaced one quartz individual, quartz having no cleavage. The quartz, however, is always bounded by ooi of mica and occurs in the mica in veinlets parallel to the cleavage plane (see figs 45, 46 and 47). Another and still more striking phenomenon is the fact that the mica scales only adjoin over extremely short distances, much shorter than would be expected considering the dimensions of the scales (figs. 45 and 46). It is obvious that the quartz has been deposited along the homogeneous mica boundaries and has replaced the mica. In fig. 46 eight quartz individuals may

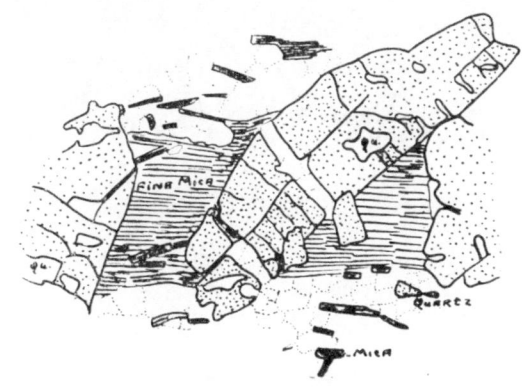

be seen, one of which (No 1) is relatively very large, all enclosing several mica scales. It is quite clear how the mica scales are only joined together by very narrow connections. A single, extremely fine scale enclosed in crystal 1 has an outline as sketched in fig. 47. There is no doubt that it has been replaced by quartz.

Fig. 48 (No 59 Kaina). Shows minute layer consisting of tourmaline and fine-scaly mica in roof rock of lens A.

Mostly this silicification is hardly perceptible, but in the country rock forming the roof of lens A it is quite obvious. This rock consists of small layers of rather coarse tourmaline, in the interstices between which is both mica and quartz (fig. 48). The dark tourmaliniferous layers alternate with white layers in

which quartz prevails, a few remnants of mica and tourmaline being left. Already under the magnifying glass it may be seen that the irregular grains of tourmaline are corroded and that quartz veinlets break through the tourmaline prisms parallel to the prismatic zone or to ooi, the ultimate result being small groups of crenelated tourmaline grains. Figs. 49 and 50 (No 59) give a microscopic picture of this tourmaline embedded in quartz. The prisms are at a small angle to the foliation (marked by mica) and it seems that rents in the tourmaline parallel to the foliation have also played a rôle. For the rest

Fig. 49,   Fig. 50.
(No. 59) Crenelated tourmaline (dotted) in quartz. Parting parallel to ooi has been sketched.

it is clear that both the homogeneous tourmaline boundaries and the most important crystallographic planes of tourmaline determine the distribution and outline of the quartz. Thus it is beyond doubt that tourmaline has been replaced by quartz.

(In pegmatitic veins similar intergrowths are seen on a larger scale. For instance photo 51 shows a tourmaline crystal corroded by quartz which was taken from a pegmatitic vein west of hill No 42, Wishikatwa. The core of the prism, which was probably of a different chemical composition than the outer zone, is preferentially replaced.)

Photo 51 (Twofold magn., polished surface, reflected light). Pegmatite vein west of hill 42 Katuba valley. Replacement of tourmaline by quartz.

The sequence of deposition resulting from this discussion is tabulated below.

1. The sericite recrystallized into fine-scaly mica.
2. Solutions deposited tourmaline,
3. while at the same time coarser mica was deposited.
4. Very coarse mica was deposited and tourmaline resorbed, particularly along the main pathway.
5. Finally silicification took place, most intensively near the main pathway of solutions.

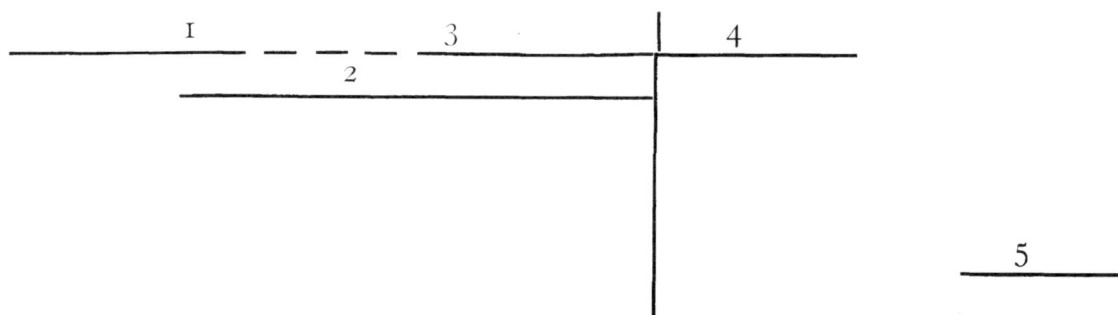

What happened between 4 and 5 will be discussed in part VA.

Another example of mineralization along fault planes is provided by Buramma Ridge. Fragments of the garnet quartzite in the northern flank of the syncline mark the fault plane and it is very interesting to observe along this fault the presence of metamorphic sediments and stanniferous veins, which normally are confined to the lower sericite phyllites below the boundary quartzite. Thus, the reverse faulting appears to have strongly reduced the impermeability of the strata and to have created another outlet for the magmatic solutions. The metamorphic rocks belong to the horizon of the garnet quartzite and consist either of quartzite or of micaceous sandstones.

The additive metamorphism is not uniform throughout this belt but most intensive around stanniferous veins or quartz blows, its intensity decreasing with increasing distance from those veins.

These metamorphic quartzites and sandstones will be described here, while examples and photos will be given of the metamorphic sandstone of Kashozo, in which also stanniferous lenses occur. However, quartzites and sandstone had already suffered from various influences before the typical "pneumatolytic" metamorphism transformed them into tourmaliniferous rocks.

For the description of the results of this phenomenon it does not matter whether these influences were exerted by load metamorphism or by dynamo metamorphism, but the result was that the sedimentary origin of quartzites and micaceous sandstones was already largely obscured before metasomatic minerals made their appearance.

Generally the aluminous cement of quartzites had been transformed into mica or sericite, while the quartz grains in the cement had been incorporated by the coarse grains. Though in some places the detrital, sedimentary character of the quartzites has been well preserved, every more or less round grain of quartz being enveloped by sericite and fine-scaly mica, in many cases the cement had almost entirely disappeared.

Photo 52 (No U 7, magn. 15x, crossed nicols). Specimen slide from boundary quartzite east of Ruitonbero. The aluminous cement has been greatly reduced and only a few scales of mica are present.

In the latter case the mica content is greatly reduced and the small scales left are found to be embraced by quartz, or else they are found along or across the grain boundaries of the quartz (photo 52, No U 7).

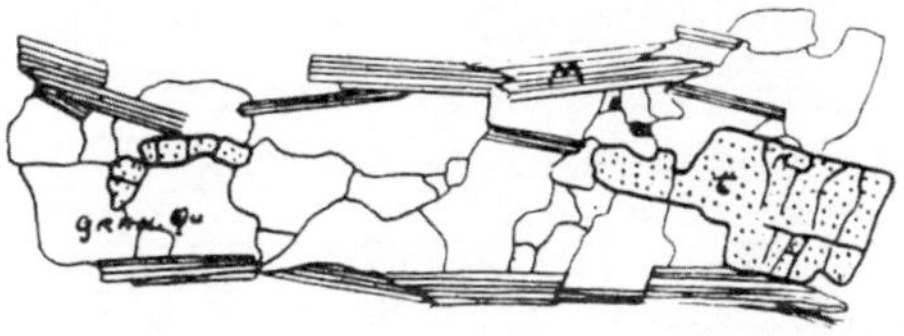

Fig. 51 (No 74). Micaceous sandstone underlying garnet quartzite on Rue-chimarra hill. Quartz (white) and tourmaline (dotted) between two streaks of mica (hatched parallel to cleavage).

The quartz grains adjoin each other along a slightly indented line and are always elongated or flattened out in the direction of the (former) stratification.

The mica appears to have been replaced by quartz and consequently is mostly bounded by its basal plane.

The sandy schists underlying the garnet quartzite on Buramma Ridge contain more mica, though here, too, part of the mica seems to have been replaced (fig. 51, No 74). In this figure some tourmaline grains are to be seen lying between two streaks of mica. The right-hand prism has developed into a hypidiomorphic shape, but nevertheless it is obvious that it is partly bounded by the arbitrary form of the quartz grains. The left-hand grains, which form one individual crystal, have been deposited along the homogeneous quartz boundaries.

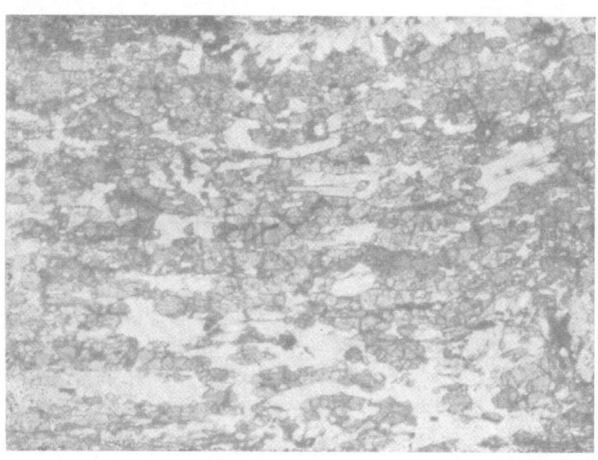

Photo 53 (No 89, magn. 13x, ordinary light). Quartz, mica and tourmaline in a slide from wall rock of vein F, Buramma.

Photo 53 is a microscopic picture of the sandy schists above the garnet quartzite and these schists also contain a fair amount of mica.

Mica is almost entirely absent in the garnet quartzite itself. Consequently the schistosity is badly discernable in these greyish, fine-grained, massive rocks. Only under the microscope (cf. fig. 53) or when tourmaline has been deposited (photo 72) can the (former) stratification be traced, for the tourmaline has been preferably deposited along the grain boundaries of the quartz and is particularly abundant as small grains along special planes, which obviously were once bedding planes.

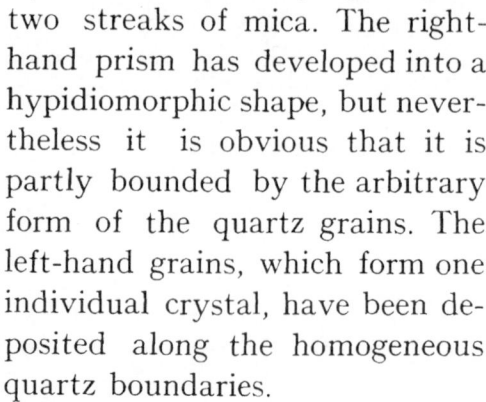

Photo 53a (No. 89, twofold magn, reflected light, polished surface). Same hand specimen as photo 53.

Fig. 52. Tourmaline (dotted), mica (hatched parallel to cleavage), quartz (white with grain boundaries) and rutile (R) in tourmalinized quartzite. Actual length of sketched part about 1½ mm.

Its grains are xenomorphic and one individual crystal often comprises several vermicular veinlets, which occur along the grain boundaries of three or four adjoining quartz grains (fig. 52).

Only in the few mica scales present does tourmaline occur in the form of more or less euhedral crystals, which invariably are situated along the edge of the scales.

Fig. 53 (No 67, Ruechimarra) does not show remnants of one single corroded tourmaline prism, for under crossed nicols it is seen at once that the separate grains are not orientated parallel, but that the whole row of these anhedral grains is parallel to the original bedding planes. These grains are not simultaneously extinguished under crossed nicols, but still their longer axis seems to lie more or less in one plane. This reveals the influence exercised by the original texture upon the course of the solutions. And not only the distribution of the tourmaline grains but even their orientation appears to be in harmony with the former stratification. To this palimpsest phenomenon the fact is due that at first sight these quartzites are classified as metamorphic sedimentary rocks, even when they are totally devoid of micaceous cement. Approaching a quartz blow the tourmaline increases in quantity; the grains

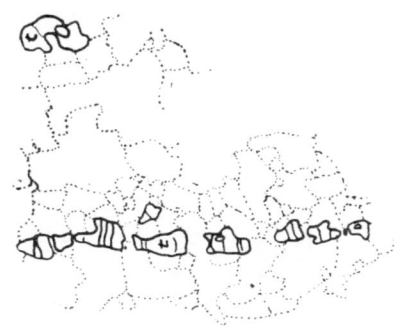

Fig. 53 (No 67a, Ruechimarra). Quartz (white with grain boundaries) and tourmaline (heavy outline and basal parting).

become coarser and black layers studded with tourmaline betray the vicinity of a former channel of the magmatic solutions.

In these black layers the tourmaline has not only been deposited along the grain boundaries, but it has also replaced entire grains of quartz. This results in a sieve-structure, a typical example of which is shown in fig. 54.

In the close vicinity of a vein or quartz blow the quartzite is sometimes entirely replaced by a mosaic of tourmaline. Often these mosaics have subsequently been replaced by quartz, though remnants are still found in the veins; undoubtedly they are remnants as every strongly tourmaliniferous layer projects a little into the

Fig. 54. (No 85) Intergrowth of quartz (white with grain boundaries) and tourmaline (heavy outline and basal parting) of which three individual crystals are present; unknown ore (black) probably an iron oxide; inclusions in quartz (fine dots); limonite (heavy dots).

vein, whilst its continuation can be traced by means of small remnants in which the palimpsest stratification is still in the same direction.

These mosaics often contain much interstitial mica, forming a rather soft rock.

Figs 55 and 56 have been drawn from slide 90 (Ruechimarra) and give an idea of the structure of such a remnant included by a quartz vein. In the first place it appears that the tourmaline crystals are not euhedral, it being very seldom that good trigonal sections are found. Between the tourmaline grains there are fairly numerous mica scales lying either parallel to one of the faces of a tourmaline grain or abutting against it with the known sawlike line.

The author considers it most probable that tourmaline is younger than mica, and as there is a fair amount of mica to be seen in the interstices between the tourmaline grains it may be assumed that the quartzite was replaced for a large part by mica prior to the deposition of tourmaline. It may also be assumed that a part of the quartz now visible was deposited by replacement of mica only after the appearance of the tourmaline.

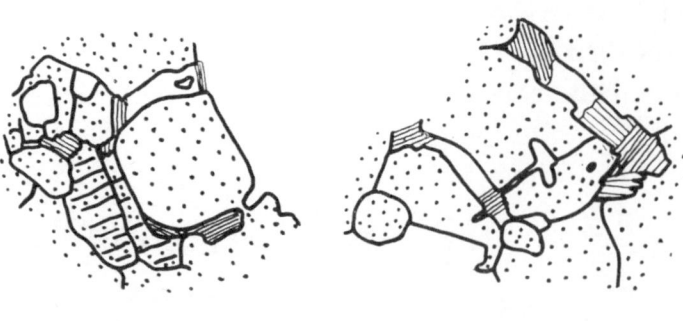

Fig. 55.   Fig. 56.

Fig. 55, 56 (No 90. Ruechimarra). diameter about 1 mm. Tourmaline (dotted), mica (hatched parallel to cleavage) and quartz (white) in remnant of tourmaline mosaic included in quartz vein.

There are, indeed, many evidences of silicification subsequent to the deposition of tourmaline. It is also visible in the quartz grains, which are of increased dimensions and are laden with crystal and gas inclusions.

This is particularly observed near the stanniferous veins and the quartz blows. Tourmaline is more or less resorbed and by this process the quartz, which was added, caused the formation of the well-known crenelated outline of the tourmaline crystals already described in previous chapters. This resorption is visible in fig. 54 near the top edge; the contact between quartz and tourmaline is formed by the basal parting or prismatic faces of the latter.

Again silicification appears from fig. 57, in which the quartz is seen to contain more or less parallel rows of inclusions. Small veinlets of quartz traverse the tourmaline prism along rents.

The final conclusions may be formulated as follows:

The additive metamorphism has been preceded by a resorption of cement and transformation of sericite in fine-scaly mica.

The sequence of deposition of the metasomatic minerals may be tabulated as follows:

Fig. 57 (No 67b, Ruechimarra). Quartz (white, grain boundaries and inclusions) filling rents in bent tourmaline (heavy dots and basal parting).

1. mica   ____I____ 1)
2. tourmaline   _____2_____
3. quartz.   ____3____

----

1) Addition only perceptible near veins and quartz blows.

In these tourmaliniferous quartzites cassiterite is sometimes present. If nothing else had happened, one would speak of an impregnation of cassiterite and its deposition would often escape notice.

However, it occurs only in the vicinity of white masses of quartz which has totally replaced the tourmaline that was formerly present. These white masses — they also contain mica and cassiterite — strongly contrast with the greyish tourmaliniferous quartzites, which still reveal their sedimentary origin.

Consequently the former pathway of the solution is not recognized in the field by the deposition of a useful mineral but by the extreme result of one single process: the silicification.

In principle the same is seen at Kaina: the vein is recognized by the resorption of tourmaline, and this resorption is due to the development of coarse mica, resulting in the formation of a white, lenticular body.

Therefore the vein space might be defined as the space in which all palimpsest phenomena have disappeared in consequence of very intensive metasomatism. Of course the results of these metasomatic processes have been particularly intensive along the main channel of the solutions. The events which have happened in the vein space will be described and discussed in the next part.

PART V

THE STANNIFEROUS DEPOSITS

CHAPTER 1

INTRODUCTION

In this part the additive processes will be described which are confined to the immediate vicinity of the main channels of the magmatic solutions. Particularly those processes will be discussed by which stanniferous veins have been formed.

In Ankole tin occurs solely in the form of cassiterite, while the only sulphide present, as far as the deposits examined are concerned, is arsenopyrite.

Only those deposits will be dealt with which the writer himself was able to study thoroughly. He had also an opportunity to see the well-known veins of Nanyankoko and Rusinga, both of which lie within the concessions of the C.A.E. Co. The visit he paid to Mwirisandu was far too short to allow of a definite idea being formed as to the genesis of that vein, but it is very probable that the mode of deposition will not differ in principle from that of the Rushenyi and Kavungo veins; this Mwirisandu deposit, however, has been thoroughly investigated by Mr Combe.

Generally speaking, the stanniferous veins are irregular, lenticular bodies. Quartz and mica are the predominant constituents and as the intergrowth generally is coarse, the veins may be called pegmatitic.

Feldspar has never been observed in these veins in Ankole, but it seems to occur in similar veins in Ruanda. Cassiterite and topaz are the most important minor constituents.

The veins are invariably accompanied by metamorphic aureoles in which tourmaline is the most conspicuous constituent.

The author cannot believe that these veins have been deposited in open fissures, nor that they have worked their way in by forcing aside the country rock.

He surmises that the aureoles have been formed by metasomatic action of hypogene solutions upon the country rock, and the formation of the pegmatitic veins is thought to be due to very intensive metasomatism along the main channel, whereby very coarse-grained minerals have been formed and all palimpsest structures, reminding one of the sedimentary host rock, have disappeared.

This assumption is supported by the following field evidences:

1.   These lenticular veins have no visible connection with the granite below. This has been proved convincingly everywhere in mining works and by surface exploration. At Kaina, for instance, tunnels have been driven underneath several lenses and though the metamorphic aureole enveloping these lenses was found to extend downwards, in some of these tunnels not a trace of the vein could be found, not even with a magnifying glass.

Moreover, several outcropping lenses have been mined. They suddenly died out in depth and trenches subsequently dug in the bottom of these lenses revealed the remarkable fact that not a trace of a channel could be found. Injection, therefore, appears to be impossible, for in that case a few coarse scales of mica along a bedding plane or a few thin lenses of quartz would reveal the passage of the vein matter.

The reverse was the case at Kabezi, where by chance a trench struck the upper part of a vein which did not extend to the surface. In the overlying tourmaliniferous schists not a trace of the vein could be found.

The limited vertical extension of the veins seems to be due to certain physico-chemical conditions causing a certain mineral to be deposited by the magmatic solutions along a short vertical height of their channels.

This view is supported by the fact that the lenses do not occur at all the different levels but are grouped together on about the same level. At Kaina, for instance, no new lenses have been encountered in the tunnels below the lenses outcropping on the hill (cf. fig. 61 part VA). Another instance is found at Rusinga, where very many lenses crop out on the hill or have been found in surface explorations, but only a few extend as far as the level of the tunnel that has been driven underneath these lenses. Therefore the optima conditions for the development of very intensive metasomatism appear to prevail at one particular level; one might wonder how it is that so many lenses are found just at the surface and so few in the tunnels, but this is explained by the fact that veins and metamorphic rocks better resist weathering than unaltered phyllites.

As they retard denudation, these veins and metamorphic aureoles consequently are most abundant near the surface.

2.   The form of the bodies is extremely irregular, both in the horizontal and in the vertical section (photo 54). They are often found to be extremely ramified and these ramifications frequently embrace large blocks of the country rock (cf. fig. 58). The foliation or schistosity of these inclusions is entirely parallel to or is the direct continuation of the schistosity of the nearest country rock. This feature cannot be explained by assuming that the inclusions have been floating in a viscous melt. The only explanation is that the vein has grown in a solid medium.

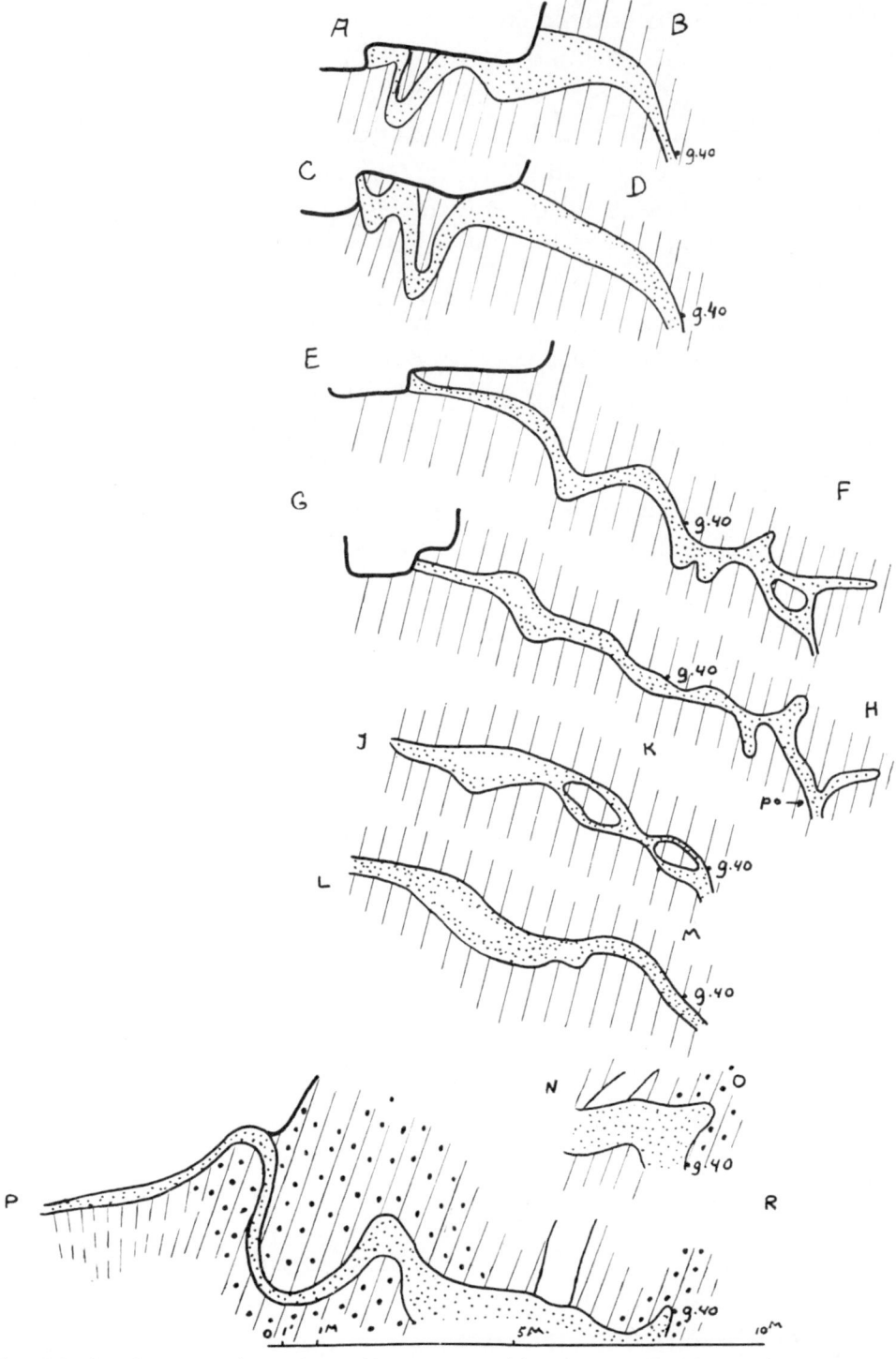

Fig. 58 Series of cross-sections of stanniferous pegmatitic vein near the Chamgash (Kavungo).
Dotted = mica predominates. White = quartz predominates. Heavy dots = tourmaline in wall-rock.
Hatched (parallel to foliation) = arenaceous phyllites of the Middle Division containing staurolite.
9.40 = point 9.4 metres above entrance to tunnel.
po = pocket.

3. The contact with the wall rock is extremely irregular and more or less vague.

Photo 54. Irregular shape of stanniferous lens D at Kaina (cf. fig. 61, Part V A). The pegmatite body has been entirely delved out and the non-stanniferous, metamorphic aureole rock has been left (the light colouring of the wall rock is due to the presence of mica).

One large lens at Kaina, for instance, consists, in the middle and near the floor, of coarse mica, whilst the upper part is silicified. The plane of contact is marked by the total disappearance of tourmaline, which in the wall rock is abundant. A close investigation reveals that the tourmaline does not disappear suddenly but that the tourmaline content gradually diminishes (photo 48 part IVB). Moreover, the white vein matter, whether mica or quartz, is seen to project [1] everywhere wedgelike into the country rock (photo 50 part IVB). This is a common feature of all stanniferous veins in Ankole.

4. The invariably lenticular veins often contain minerals inherited from the country rock and left in the vein after replacement of the country rock.

Thus cyanite, for instance, is found in micaceous quartz veins wherever this mineral is a component of the country rock. The distribution of cyanite has been discussed in Part III and consequently the area in which cyanitiferous veins are found is known to the reader. On the slope of the Ruitonbero garnets were found in a mica vein that was only partly silicified. This vein occurred in the garnet phyllites and the garnets in the vein were exactly similar to those in the phyllites. Garnets, however, have never been observed in veins occurring in non-garnetiferous phyllites or quartzites.

The texture of the fine-scaly mica present in the veins often suggests that this mica is a recrystallized remnant of the country rock, especially where it occurs along the walls of the veins (cf. photos 55 and 56).

Photo 55. (Twofold magn., polished surface, reflected light.) A hand-specimen of stanniferous lens occurring below boundary quartzite at Kabezi (cf. fig. 79, Part V B). Mica, quartz and cassiterite (black, top right-hand corner) are present. The texture of this pegmatite suggests that the mica originated from recrystallization of sericite phyllite.

5. Often the lenses lie practically horizontal (cf. fig. 66 part VA Kaina and

[1] merely used for the sake of easy description.

fig. 58 Chamgash) and it is most unlikely that these have been open fissures.

6.   There is no evidence of deflection of strata along the veins. They abut

Photo 56. (Twofold magn., polished surface, reflected light.) Hand specimen showing the contact at floor of lens D, Kaina, with tourmaline abundant in the wall-rock. The white vein matter consists, away from the contact, of coarse mica (white; right-hand edge), but near the contact it is rather fine-scaly and its texture suggests that it originated from recrystallization of phyllite.

against the lenses and disappear, but never sweep around them. Layers studded

with tourmaline often extend a short distance into the vein, and patches consisting of strongly corroded tourmalines reveal their former extension in the vein space.

Finally, it is proved that the various vein minerals have also replaced each other and that they have not been deposited simultaneously.

In hand specimens it is clearly seen that quartz has replaced older minerals, such as tourmaline and cassiterite; it fills fissures and forms veinlets in broken tourmaline (cf. photo 51 part IVB, and photo 57 or in cassiterite (cf. photo 58), whilst it has undoubtedly replaced mica too.

Consequently this implies that the presence of quartz is unfavourable for the value of the veins. Results of sampling are in harmony with the above-mentioned de-

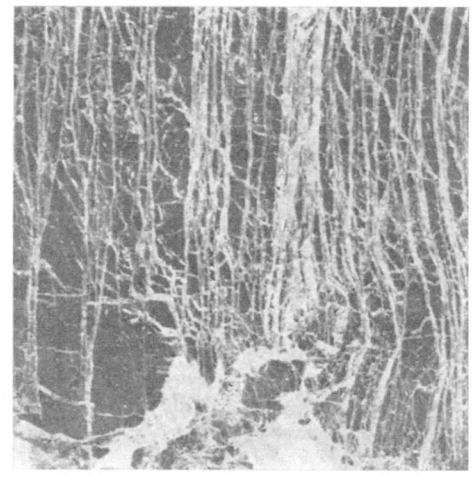

Photo 57. (Twofold magn., polished surface, reflected light.) Shows coarse, cracked crystal of dark tourmaline from pegmatite vein west of hill 42, Katuba valley. The rents are mostly filled with quartz, but a little mica is also present.

duction. Careful bulk sampling invariably revealed the fact that where quartz prevails over mica the tin content is much smaller than where mica prevails over quartz. Naturally, in order to be comparable, the samples should be taken in the closest possible vicinity to each other.

At Kaina, to take an actual example, the big lens which has been previously mentioned and of which only the roof is silicified, contained about 1.5 % of Sn metal, while the silicified roof contained less than 0.2 %.

The optima conditions for the deposition of several minerals appear to overlap, as in the case of tourmaline, cassiterite and coarse mica (younger than tourmaline).

This could be noticed at Kaina, where on Kaina hill these minerals occur on about the same level and tourmaline is abundantly present in the wallrock. Coarse mica and cassiterite are found together and often do not reach the upper part of a lens, though they have been proved to exist at deeper levels. In this case the outcrop consists of quartz and fine-scaly mica.

On the slope of the Nobugamba a vertical row of lenses is exposed, lying

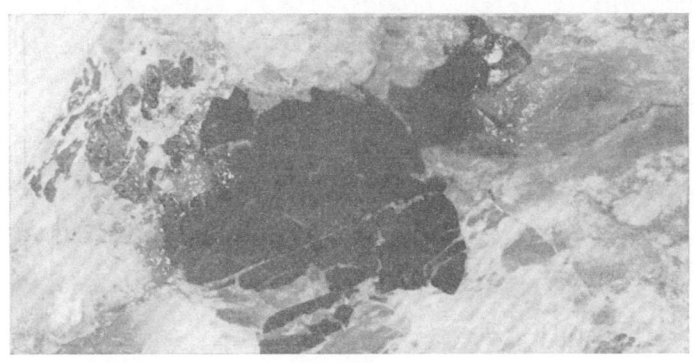

Photo 58 (Twofold magn., polished surface, reflected light). Hand specimen from vein G, Buramma Ridge (cf. fig. 79, Part V B) showing replacement of cassiterite by quartz, the latter filling rents in the coarse, broken cassiterite grain embedded in quartz having suffered from stress.

parallel with and at some distance from an important fault plane (cf. fig. 59, part VA). The lower lens contains cassiterite, a little coarse mica and quartz. A little tourmaline is present in the wall rock. The next higher lens is quite similar to the first but differs in that cassiterite is absent. Still higher the lenses contain only very fine-scaly mica and quartz and in the wall rock tourmaline needles are almost absent. On the top the lenses merely consist of quartz with but very few fine scales of mica. No tourmaline can be traced in the country rock. The vertical interval between the lowest lens and the top one is about 120 metres (cf. fig. 59, part VA).

This succession, in a general sense, also holds good when proceeding from the sericite phyllites to the garnet phyllites away from the granite. In the sericite phyllites pegmatitic veins are very numerous and often contain coarse mica and cassiterite. The frequency of cassiterite occurrences in the lowest Karagwe-Ankolian horizons is evident from a glance at map 3. Though on the whole the sericite phyllites are intensely mineralized, the pegmatitic veins contained therein are mostly of small dimensions, except where a fault is present, as at Kaina. These aluminous phyllites are much more mineralized than the garnet phyllites. The latter belt mostly contains only barren quartz blows (photo 59), except where it abuts against the granite. Above the Ihunga quartzite even quartz blows become scarce, at least where the normal succession of strata lies between them and the granite.

But even where these higher horizons of the Karagwe-Ankolian system are in direct contact with the granite these belts seem to offer poor prospects, for generally the quartz lenses found there are barren too.

The ore-geologist, therefore, will be more favourably impressed –– as regards the possibilities for tin — by the sericite phyllites than by the garnet phyllites and the stratigraphically higher horizons. Nevertheless these belts contain some promising veins! The tectonics may account for their deposition, as is illustrated by the distribution of stanniferous veins on Buramma Ridge (cf. fig. 79 part VB); on this ridge many veins occur along a reverse fault (cf. section 6), the throw of which may be small. These deposits which occur in or near the normally barren garnet quartzite are even more promising than the numerous but very small veins found below the boundary quartzite. The influence of faults has already been demonstrated in connection with the occurrence of tourmaline in the garnet phyllities on hill A (cf. part IVB and

Photo 59. Barren quartz blow in upper sericite phyllites above the boundary quartzite on Ruechimarra Hill.

map 5), whilst the occurrence of tourmaline north of the Chamiombu in the core of a strongly compressed syncline is undoubtedly due to tectonical causes.

As a result, the first impression obtained by the ore-geologist is erroneous, for the composition of the country rock and the distance from the granite are not the only determining factors for the deposition of veins, since faults also play a highly important rôle.

Some veins or lenses are not comprised under the term metasomatic veins, as their mode of formation is doubtful. Two different types may be recognized. One type is exclusively found west of the Omutarraz and on the Shonobutondo in the sericite phyllites, thus in the sharp

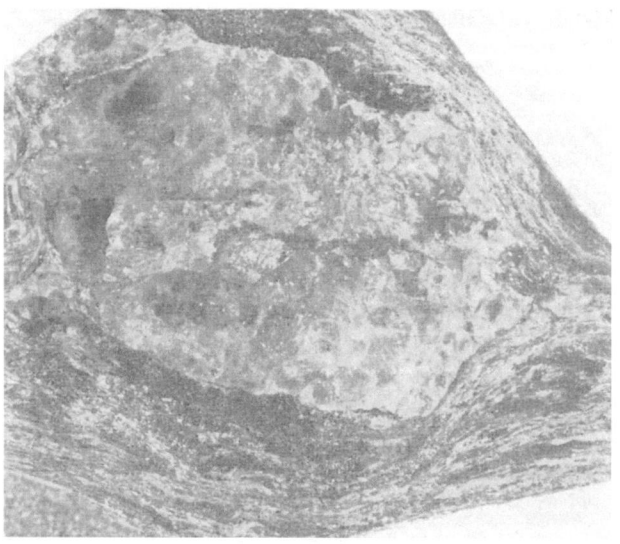

Photo 60 (Twofold magn., polished surface, reflected light).
Fist-like lens of quartz occuring in tourmaliniferous sericite phyllites on Kavusanammi West.

bend of the latter around the granite. This may account for their formation, as compression might have forced the strata apart; indeed, the strata are deflected and sweep around these veins. They are very long in proportion to their thickness and have straight and very sharp walls. The filling is uniform and consists of mica and quartz, while topaz is sometimes present. In thickness they range from 1 foot to a few millimetres and one of them is shown on photo 45, part IVB. Presumably they are younger than the deposition of tourmaline, as the latter mineral occurs abundantly in the wall rock but never in the veins.

The other type is only found to occur in the disturbed zone of the Kavusa-nammi up to the Kakenenne. These veins occur both in metamorphic (cf. photo 60) and in unaltered sericite phyllites, and consist of small fistlike bodies of a bluish, glassy quartz. Their formation might be related to the tectonics of the zone in which they occur (cf. maps 4 and 5 and fig. 22 part III).

A. KAINA

CHAPTER 2

GENERAL DESCRIPTION

The Kaina deposits (photo 61) are the most important deposits known to exist wthin the area under consideration, though up to 1931 they yielded only 12 metric tons of tin metal.

They are in genetic relation to a normal fault, along which a quartz reef

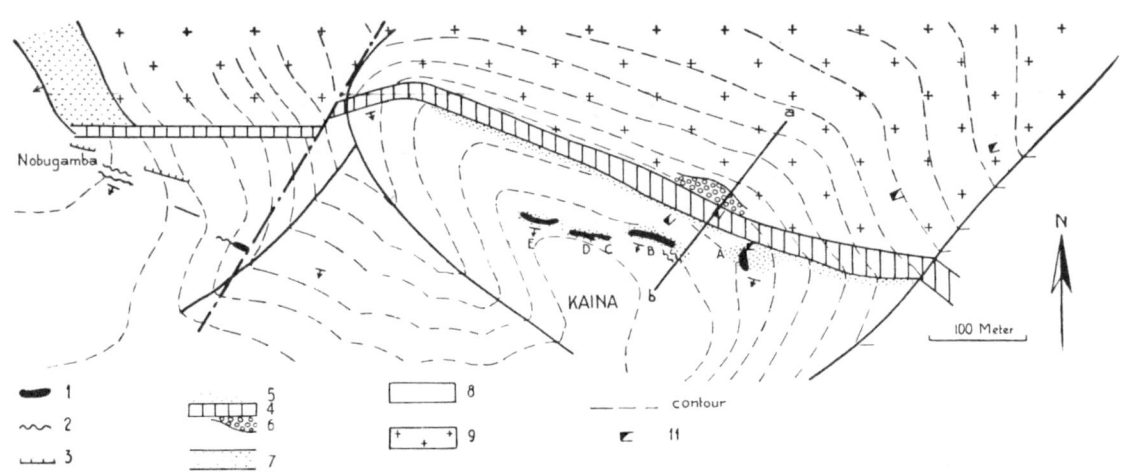

Fig. 61 Sketch map of Kaina Hill.

1 — stanniferous vein.	6 — metamorphic phyllite with lenticular texture.
2 — barren quartz mica vein.	7 — quartzite.
3 — barren quartz lens.	8 — phyllite.
4 — quartz reef.	9 — granite.
5 — metamorphic phyllite.	11 — adit.

has been formed; this fault has already been discussed in part III (cf. fig. 27 part III), whilst the metamorphism in the sediments has been described in chapter 6, part IVB.

In the phyllites on Kaina hill five lenses occur almost in one line. The strike of these lenticular veins coincides with the strike of the sediments, but the dip of the lenses is variable, being either steep southerly, in which case the lenses are conformably intercalated in the sediments, or gentle northerly, thus hading against the dip of the phyllites. The situation of the veins is seen in fig. 59, a sketch map of Kaina hill.

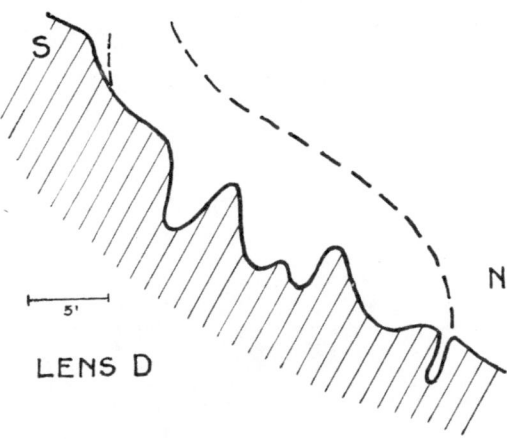

The lenticular bodies all have a very irregular shape and the contact with the wall rock is also irregular.

They cut across the strata or make a sudden bend and run parallel to the schistosity plane of the country rock. In thickness they are extremely variable. Figs. 60 and 61 show two cross sections through two different lenses

LENS D

Fig. 60 Cross-section. eastern end of lens E. Wall rock hatched parallel to foliation. Crosses indicate vein matter consisting chiefly of mica. White = quartz predominates.

Fig. 61. Cross section lens D (cf. photo 54, part V). Wall rock hatched parallel to foliation. Former shape of stanniferous lens (now delved out) shown by dotted line.

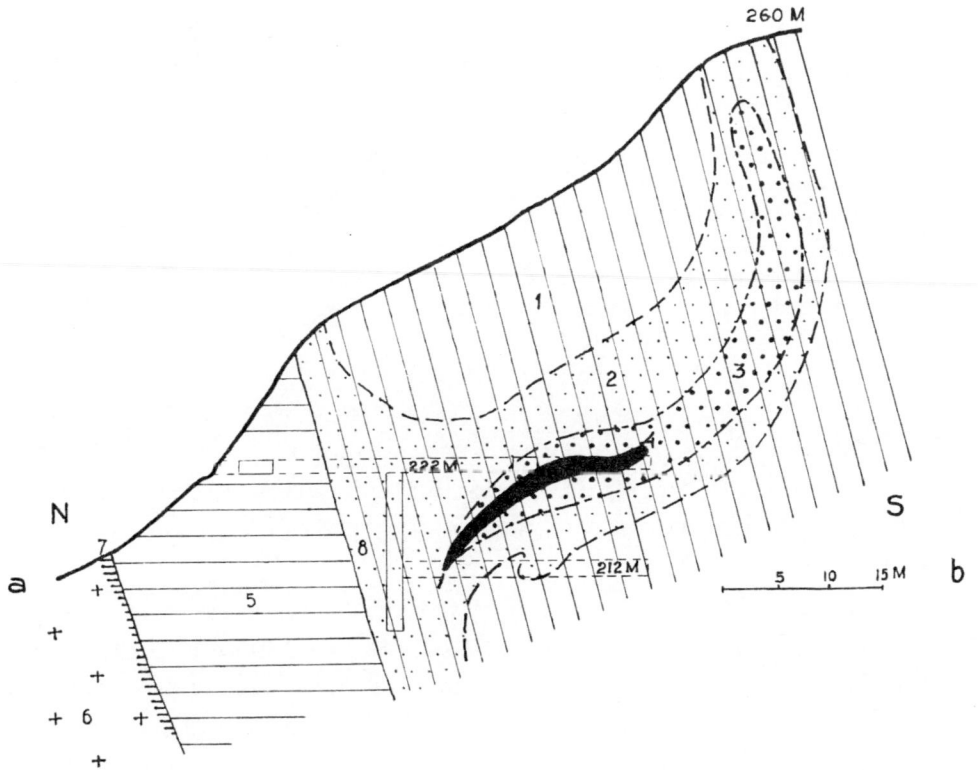

Fig. 62. Cross section lens A through shaft and sampling crosscut (section a—b fig. 59).

1. Blue sericite phyllite.

2. } Metamorphic aureole { yellow tourmaliniferous mica schists.
3. } { green tourmaline schists.

4. Pegmatitic vein, floor of which consists of aggregate of coarse mica.

Heights in metres above Kaina House.

5. Quartz reef.
6. Granite.
7. Lenticular schists (cf. photos 62 and 63).
8. Lenticular wall-rock (cf. photo 64).

and it is to be seen that the schistosity planes of the country rock do not sweep around them. Fig. 62 is a cross section through the most important lens (lens A) and shows the vein approaching the big quartz reef in depth, which fact is also shown by two other lenses.

Fig. 63 finally gives a picture of the exposures in the quarry on the eastern flank of Kaina Hill. Here the same feature is noticed. Moreover, it appears from fig. 59 that the outcrops of all lenses lie at about the same distance from

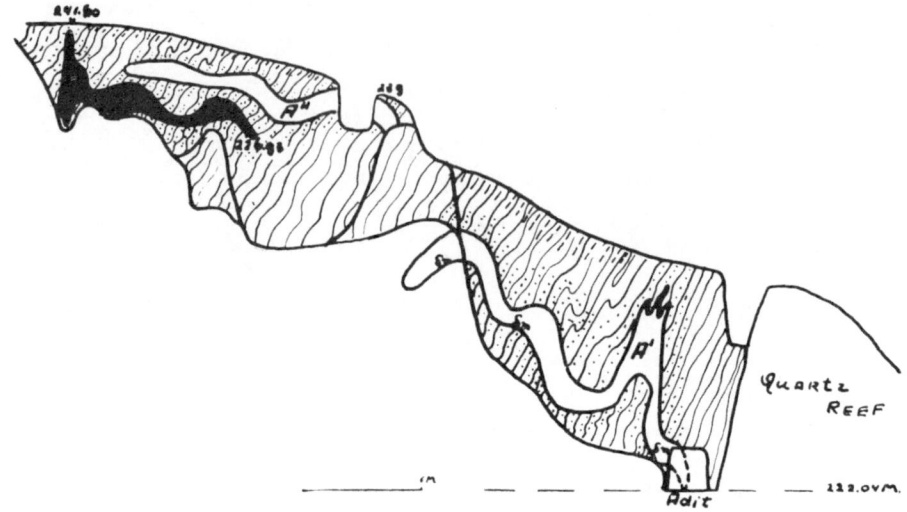

Fig. 63. Exposures in quarry near adit on eastern flank of Kaina Hill (cf. fig. 59).
        black = outcrop of lens A.
        A' = stanniferous quartz vein.
        A" = barren quartz vein.
        hatched (parallel to foliation) = schists and phyllites.
        dotted = green tourmaline schists.
        - - - - = blue sericite phyllites.
        Heights in metres above Kaina House.

the quartz reef, which is accompanied by a similar aureole as the stanniferous deposits. Undoubtedly the stanniferous deposits are in genetic relation to the fault along which the quartz reef has been formed. Therefore, this quartz reef will first be described in detail.

It is about 3 km long and the vertical height over which its outcrop is visible is nearly 400 metres, while its width, though variable, is approximately 20 metres. It runs in a straight line from one end of the boundary quartzite on Nobugamba hill to the other end near the Kabale road, thus connecting two ends of the boundary quartzite (cf. photos 3 and 7, part I, and photo 61 of this part) which have been separated by faulting. As the quartz reef marks the abnormal contact between granite and phyllites it is certain that it has been deposited along the fault plane. At the southern wall it is accompanied by a thin belt of metamorphic phyllites ("5", fig. 59), of which the intensity of metamorphism varies considerably. At the northern wall small remnants of a peculiar metamorphic schist with lenticular texture are wedged in between the granite and the quartz reef ("6", fig. 59; cf. fig. 27 part III). As probably the fault plane has not been absolutely flat and the quartz

reef has not followed every outward bend of this plane, small remnants of those phyllites which have suffered directly from the crushing influence produced by the relative movement of the two blocks along the plane of contact have escaped silicification and are now found as wedges between granite and quartz reef. These phyllites have decidedly suffered from dynamo-metamorphism. Microscopical investigation proves this rock to be built up by coarse eyes of quartz enveloped by very fine-scaly mica, which clearly sweeps

Photo 61, Kaina Hill. Showing surface exploration of stanniferous veins in phyllite above quartz reef. Waste from tunnel is seen on left slope. Under the big quartz reef is granite. In the foreground the second quartz vein (cf. fig. 27. Part III).

around the eyes or is seen to be wedged in between two adjoining eyes (photo 62). The eyes are not intensely crushed, but some have been broken into a few parts.

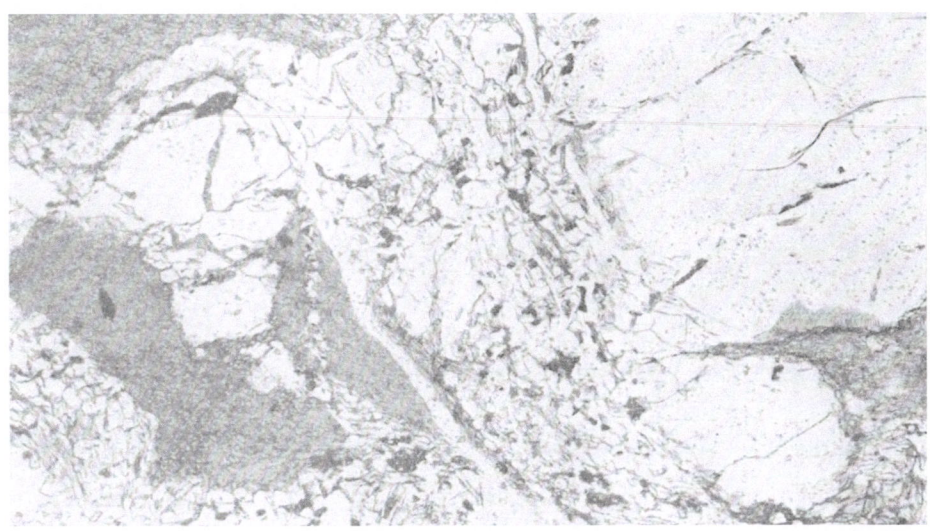

Photo 62 (No 28, Kaina, magn. 22x, ordinary light). Eyes of quartz and crushed sericite phyllites in places wedged-in between the eyes. Northern wall rock of quartz reef between this and the granite.

There must undoubtedly have been a rolling movement in this phyllite, for it is only by such a movement that the peculiar texture shown in photo 62 can be explained. The fact that the quartz eyes are not finely crushed may be due to two circumstances: in the first place they were embedded in a soft and yielding medium, and secondly it is possible that recrystallization continued during the deformation; probably the latter was just as important as the former.

The ground mass in which those quartz eyes are embedded consists of an equigranular aggregate of very fine-scaly mica (see photo 63 which undoubtedly suffered from crush; convincing proof is the increasing fineness of the mica scales

Photo 63 (No 21, magn. 30x, ordinary light). Same rock as photo 62, showing aggregate of mica and quartz eye. Mica finely crushed in vicinity of eye near bottom edge of photo.

in the vicinity of the quartz eyes. The larger porphyroblasts of mica, however, do not show signs of bending. Here and there this ground mass of mica is partly replaced by quartz, which is present in very fine grains lying in the interstices between the mica scales. As a result an extremely fine intergrowth of quartz and mica is formed, occurring in patches in the ground mass (cf. photo 62). The quartz, showing a preference for the homogeneous mica contacts, is bounded by 001 of mica and envelops sometimes very fine remnants of mica (fig. 64). The quartz also breaks through the mica-porphyroblasts in the very thin veinlets (fig. 65).

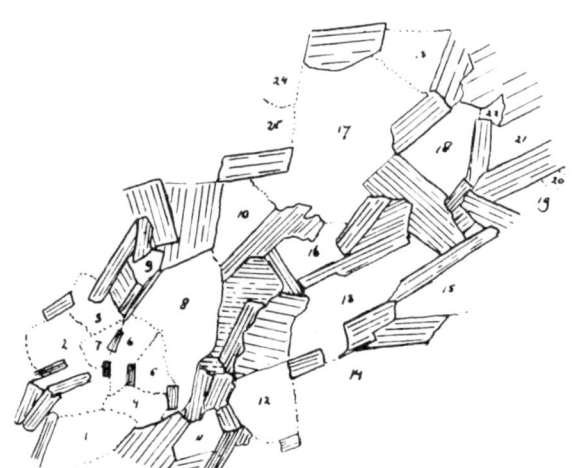

Fig. 64 (No 28 Kaina). Fine-grained intergrowth of quartz and mica, also visible in photo 62.

There is, therefore, no doubt that silicification has taken place after the phyllite had been crushed and the mica porphyroblasts had been formed, and these silicified parts show no traces of cataclasis.

Undoubtedly the coarle quartz eyes are older than or at least have been simultaneously formed with the movement that has taken place along the fault plane. The fine-grained intergrowth between quartz and mica, however, has been formed after the cessation of the main movement. A few tourmaline crystals have been deposited in this rock.

The formation of the quartz reef probably coincided with the subsequent silification observed in the metamorphic wall rock, for no traces of movement or crush can be found in the reef quartz. Its metasomatic mode of formation is revealed by slides taken from the southern contact with the sediments.

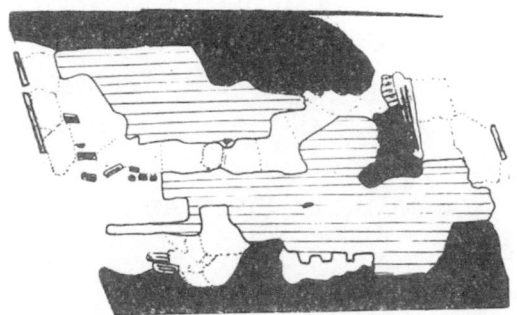

Fig. 65 (No 28). Small veinlet of quartz (white with grain boundaries) in mica porphyroblast (hatched parallel to cleavage); fine-grained intergrowth of quartz and mica (black).

There is a rather great difference between the texture of the schists at the southern wall and of those at the northern wall and this difference suggests that once a link existed, the disappearance of which is undoubtedly due to the deposition of the quartz reef, quartz having replaced about 20 metres of strongly crushed schists.

However, the schists forming the southern wall are still abnormal in texture. These yellow tourmaliniferous mica schists show a distinctly lenticular texture (photo 66). Strain-slip cleavage is obvious. Tourmaline crystals are broken and quartz fills the rents.

These tourmalines seem to have been partly replaced by quartz; they display the well-known crenelated forms and their contact with quartz is parallel to the prismatic zone and basal parting of the tourmaline.

In the bottom right-hand corner of the photo it may be seen

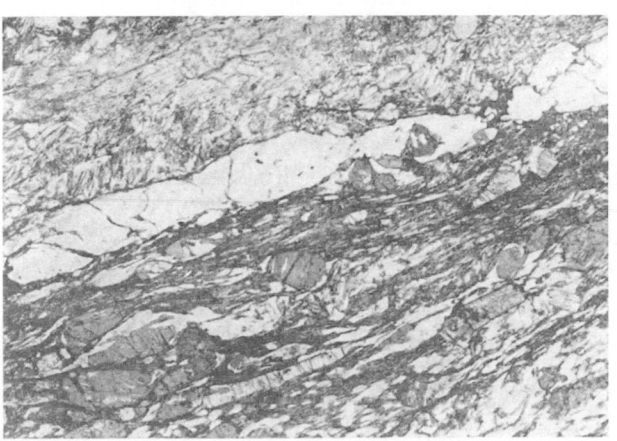

Photo 64 (No 60 from shaft of Kaina, of "8" fig. 62, magn. 17x, ordinary light). Quartz lenses, tourmaline and mica.

how the mica is also being partly replaced by fine-grained quartz of interstitial occurrence. Undoubtedly the vein-forming solutions ascended along such schistosity planes as seen in photo 64, and it is indeed highly probable that a genetic relation exists between the stanniferous lenses and the faulting. A microscopical examination of the wall rock of the reef leads to the conclusion that the metamorphism probably began during the normal faulting.

No cassiterite having been found in the quartz reef, attention may again be drawn to the stanniferous bodies. For the greater part these had been despoiled of their richer offshoots by previous prospectors, who left the unpayable quartziferous portions. From these lenses a good idea could be obtained as to their form and relation to the country rock, but they gave very little information in regard to their average mineral composition and tin content. Anyhow it was still possible to determine the minimum value of these veins.

The upper part of vein E is very rich in quartz and relatively poor in mica, which latter is coarsely intergrown with the quartz. The mica is present as large but thin folia. The width at the top is 1 metre. Where the lens thins out (cf. fig. 61) the percentage of mica increases, as also the cassiterite. The narrow feeding channel (10—15 cm wide) exposed on the east and west sides consists almost entirely of a coarse aggregate of mica intergrown with very coarse xenomorphic crystals of cassiterite with an adamantine lustre and greyish-brown colour. The radial aggregates of mica are mostly implanted perpendicularly to the walls, and wedge-like or truncated pyramids consisting of mica intergrown in two or more directions may be detached; the interior of these pyramids sometimes consists of cassiterite.[1] The feeding channel embraces a triangular piece of country rock (cf. fig. 61), the foliation of which runs in the direct continuation of the foliation of the wall rock. Photo 65 gives a view of the western side of lens E and fig. 61 of the eastern side; particularly in the latter it is clearly seen that the phyllites have not been bent. A bulk sample taken from the top part of the vein gave 0.4 % Sn, whilst another bulk sample (volume 1.3 m³) from the whole of the section shown in photo 65 yielded 0.52 % Sn. Ten per cent of the total amount of cassiterite washed out of

Photo 65. Arrow-head indicates dip of wall rock. View of western end of lens E exposed in winze. 1 = phyllites; 2 = vein.

this latter sample was found in a pocket near the floor, the volume of which was about 3 cubic decimetres! This gives clear evidence of the irregular distribution of the cassiterite.

The relation between mica content and cassiterite value was also demonstrated by lens C, the western part of which consisted chiefly of mica, whilst its eastern part was mainly quartz; a bulk sample from the former part gave 0.42 % Sn-metal, whereas the quartzose part yielded only 0.265 %.

Lens A, of which a cross section is shown in fig. 62, was the only one which had not been high-graded. It was thoroughly investigated by means of drives and cross-cuts, but unfortunately it was definitely established that it did not extend downwards and consequently exploration was discontinued.

---

1)   and of topaz (cf. photo 71).

Fig. 66 gives a section over a sampling cross-cut through lens A, which yielded 18 tons of rock. The average tin content was 1.44 %, though in the silicified roof part only there was less than 0.2 %.

The dip was exceedingly flat here and the general appearance greatly supports the hypothesis of metasomatic replacement. The floor part of the vein consisted of coarse mica, showing the same intergrowth as has been previously described in connection with lens E.

Near the floor large pockets of 100 to 200 lb. cassiterite were found, intimately intergrown with mica, so that no large blocks of the cassiterite could be extracted. After blasting out the rock about 40 % of the total cassiterite was recovered by handpicking on the $^3/_8''$ sieve, only a small portion (about 10 %) having a diameter larger than $1''$.

Fig. 66. Section of lens A exposed in sampling cross-cut.

The part of the rock which after crushing [1]) passed through a 20-mesh sieve (about 30 % of the total volume) could be dumped as waste, the content of cassiterite being negligible. This proves that the intergrowth is not very fine.

Photo 66 is a two-fold magnification (under reflected light) of a specimen from the floor of vein A (cf. also photo 49, part IV B). It depicts exactly the general occurrence of cassiterite in these veins.

Near the floor intergrown with mica also large pockets of topaz occur, some of which are a few cubic feet in volume, whilst all of

Photo 66 (two-fold magn., reflected light, polished surface). Contact at floor of lens A Kaina. Near top edge coarse crystal of cassiterite intergrown with aggregate of coarse mica. Tourmaline abundant in wall-rock.

them are almost entirely devoid of cassiterrite. Photo 67 shows topaz, cassiterite and mica all near the floor. Topaz also occurs in the micaceous feeding channel of lens E, but in the quartziferous roof of vein A it is wholly absent.

Near the roof vein A is entirely devoid of coarse mica. Only some tourmaline crystals are to be seen in the barren quartz and these are certainly remnants of the roof rock that escaped replacement; it can clearly be seen how layers composed entirely of tourmaline project into the quartz and die out, the tourmaline grains becoming sparser and sparser. Coarse grains of cassiterite are seldom found in the roof quartz.

---

1) "graded" crushing.

Stheeman, Geology.

The rare occurrence of mica, cassiterite and topaz in the quartz-rich parts of the vein is a remarkable feature needing explanation. Microscopical investigations will show that the rare occurrence of these minerals is due to replacement by quartz subsequent to their deposition. Before proceeding to the microscopical examination of the ores, of which the principal sequence of deposition is the most important object, another feature must still be mentioned, namely the invariable occurrence of very few arsenopyrites with the cassiterite; for every 20 lbs. of cassiterite 5—10 small crystals of arsenopyrite were washed out.

In the quartz of the quartz reef, in the metamorphic phyllite and also in the quartz of the lenses, peculiar cavities were detected, which, judging from their shape, originated from weathering of arsenopyrite.

Photo 67. (Two-fold magn., polished surface, reflected light). Contact at floor of lens E Kaina. Tourmaline abundant in wall-rock. On left two coarse crystals of topaz. In middle coarse grains of cassiterite. Top edge topaz and cassiterite intergrown with coarse mica.

CHAPTER 3

MICROSCOPICAL INVESTIGATIONS

A: Cassiterite.

Under the microscope the cassiterite was found to be faintly pleochroic from impure white to faint red, this pleochroism being more easily perceptible in the darker coloured places. The variations in colour may be seen from photo 68, a four-fold magnification of a slide (No 82 B, Ruechimarra) under reflected light. The darker colour extends particularly along the twinning planes and it is clear that not only these planes but also the grain boundaries and the cleavage determine the location of those darker patches. It is possible, therefore, that this colouring is due to infiltration of foreign matter, subsequent to the formation of the cassiterite itself. The same photo also clearly shows a perfect cleavage parallel to 001 and a less perfect cleavage parallel to 111. Further, there is the peculiar twinning which also characterises rutile, the twinning plane being probably 101. For the rest the cassiterite shows the usual high relief and high birefringence.

This description holds good for the cassiterite of the Kaina, Ruechimarra and Kashozo deposits.

Photo v. Werkhoven.

Photo 68 (fourfold magn., reflected light). Slide No 82b Ruechimarra, showing colouring of cassiterite and intergrowth with mica (cf. fig. 96).

A remarkable phenomenon observed exclusively in the cassiterite of these three deposits is the presence of very fine lamellae occurring in two different directions. These are shown in photo 69, in which a triplet is to be seen. The ptical orientation of the lamellae corresponds to that of one of the triplets, and consequently a complicated sort of penetration triplet isformed.

Photo 69 also shows a quartz veinlet bounded entirely by the main crystallographic directions of the cassiterite crystals; the influence of twinning plane and cleavage is obvious. From photo 76 Part V B it may be seen how the abundance of twinning lamellae may give cassiterite an appearance similar to that of calcite.

Photo 69 (No 38 Kashozo, magn. 14x, crossed nicols). Penetration triplet of cassiterite. Near bottom right-hand corner few tourmaline prisms embraced by cassiterite. Top right-hand corner quartz.

B: Cassiterite and Mica.

In Part IV B, chapter 4, two sketches of Kaina slides were discussed, and the conclusion arrived at was that mica had been replaced by cassiterite and that the latter in turn had been replaced by quartz. In consequence the cassiterite was given a peculiar outline, as evident in fig. 67 (No 13), where the cassiterite is seen to penetrate between two differently orientated mica booklets. According to the writer's experience, the fact that cracks often occur in the cassiterite in the receding corners has not the least significance; in a mineral with such a high relief as cassiterite innumerable cracks are always found, whereas in quartz for instance very few cracks indeed are visible.

Since these cracks often form the extension of mica plates in the cassiterite, one might be inclined to assume that the latter has been partly replaced by mica, deposited along cracks. The formation of these cracks, however, might be due to stress subsequent to the deposition of both cassiterite and mica. Probably the soft mica gave way and the unsupported wedges of cassiterite became cracked.

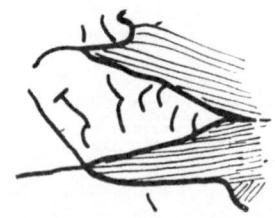

The author considers this hypothesis to be probably the right explanation, for the occurrence of cassiterite along the homogeneous mica boundaries is a strong argument in favour of the assumption that cassiterite is younger than mica.

Fig. 67. Slide No 13, Kaina, showing cassiterite (heavy outline) with cracks and mica.

When examining a slide it must always be borne in mind that it is a two-dimensional plane. In fig. 68 (No 16), for instance, a veinlet of mica seems to separate two cassiterite individuals orientated parallel.

That this veinlet of mica is younger than cassiterite cannot be proved, for the slide may arbitrarily cut the heterogeneous contact between mica and cassiterite. A conclusion can only be drawn when one mineral is always observed to form veinlets in another mineral and the reverse case is never met with. As the cassiterite is seen to project in the mica parallel to the cleavage of the latter, and the heterogeneous contact is always parallel to the basal plane of mica, it is quite probable that here again the cassiterite is younger than mica.

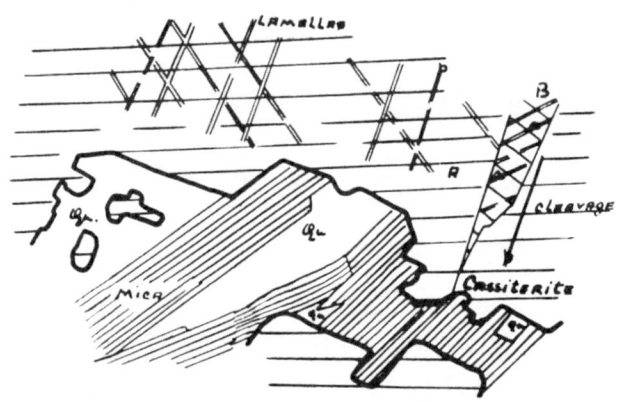

Fig. 68. (No 16a Kaina). Cassiterite with lamellae and cleavage, quartz (Qu — white) and mica.

Suppose, for instance, that a slide had been cut perpendicular to the plane of fig. 67 (or of figs. 31 and 32 part IV B) and just intersecting the top of the wedges of mica, one would obtain a slide in which veinlets of mica are found in cassiterite. The heterogeneous boundary, however, would probably be parallel to the basal plane of mica, as cassiterite is undoubtedly younger than the latter. This is actually what is shown by fig. 68, and a similar case is observed in fig. 69. In this particular case, however, quartz has partly replaced the mica, and the outline of the quartz veinlet still reveals the former presence of mica. Quartz is the youngest constituent which has been deposited along a heterogeneous contact, preferably replacing mica. It is also observed in fig. 68, where it occurs along the homogeneous and heterogeneous mica boundaries.

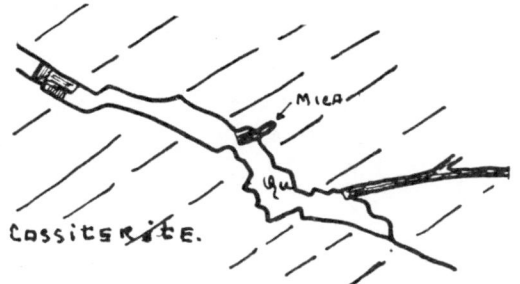

Fig. 69 (No 12 Kaina). Intergrowth of cassiterite and mica, latter partly replaced by quartz (white).

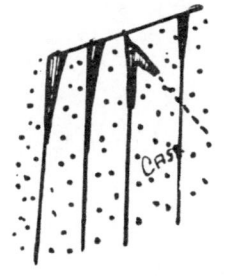

Fig. 70. Mica (hatched) along cleavage planes in cassiterite (dotted).

A younger mica, however, may occur, though in small scales and rather sporadically, for sometimes mica fills cracks in cassiterite, and then the mica scales are bound by the arbitrary form of the crack. Elsewhere a very fine scaly mica occurs along the cleavage planes of cassiterite (fig. 70).

C: Cassiterite and Quartz.

In Part IV B, chapter 4, the relation was shown between cassiterite and quartz with reference to fig. 32. As a matter of fact this very same relation is always observable, so that there is no need to go into this further.

Moreover in the foregoing it has been pointed out several times that quartz is younger than cassiterite, and it may be taken as an established fact that quartz is the youngest constituent of these veins (apart from arsenopyrite). It appears to have replaced both mica and cassiterite, which fact explains the relation between the cassiterite value and the ratio of mica to quartz shown by sampling.

Remnants of cassiterite grains of all dimensions are found in the quartz, but sampling revealed the fact that the average cassiterite grain in quartz is smaller than in mica (about 4 mm in quartz against about 9 mm in mica).

A peculiar phenomenon, however, occurring in slides cut from grains of cassiterite taken from near the roof in the quartziferous upper part of vein A must still be described. It is the presence of quartz inclusions of a roughly trigonal outline in the cassiterite, which inclusions are especially abundant near the grain boundary of the cassiterite (cf. figs. 71 to 76). They give the cassiterite a peculiar indented outline. In the quartz forming the inclusions there are always rows of very minute inclusions. Often two adjacent inclusions are extinguished simultaneously under crossed nicols, proving that they belong to the same crystal individual; this has been shown in figs. 72, 73 and 74. Furthermore, it is sometimes seen that an inclusion is extinguished simultaneously with a quartz grain lying close up against the grain boundary of the cassiterite (fig. 71 — grain a' and inclusion a), or else that two or more trigonal inclusions adjoin each other along one side (fig. 73) and are filled with one quartz crystal; the same is seen in fig. 74, where two roughly trigonal inclusions of mica are contained in the cassiterite.

Is it possible that these inclusions are the remnants of one large and partly replaced quartz crystal? As the faces of the inclusions are never parallel to the main optical directions of the quartz, this is improbable.

For the same reason these inclu-

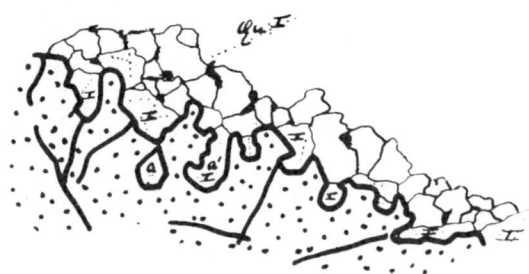

Fig. 71 (No 25 Kaina). Cassiterite (dotted) with cracks; vein quartz with inclusions (I) and clear small grains of quartz originating from crush. Quartz grains (a, a') extinguished simultaneously under crossed nicols.

Fig. 72 (No 40). Vein quartz with rows of inclusions and cassiterite (dotted).

sions cannot have been formed by replacement of cassiterite by quartz developing euhedral crystals. Moreover, several inclusions are filled with but one quartz individual. The only explanation seems to be that a trigonal mineral, embraced but not replaced by cassiterite, has subsequently been replaced by quartz. This mineral undoubtedly was tourmaline.

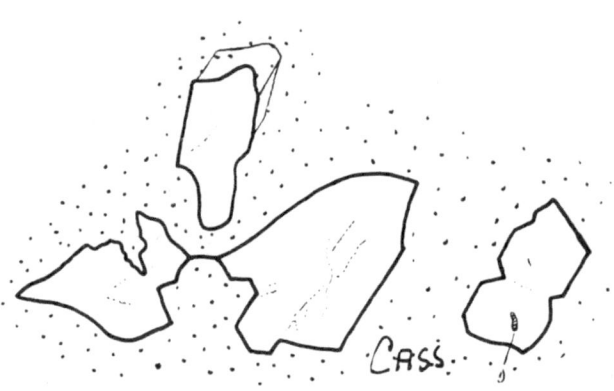

Fig. 73 (No 18). Inclusions of quartz in cassiterite (dotted). 1) = remnant presumably consisting of tourmaline.

Indeed, there are many evidences that tourmaline has never been replaced but always embraced by cassiterite. Photo 70 gives evidence of tourmaline, enclosed in cassiterite, at the same time time showing these quartz inclusions (cf. also fig. 83 and photo 75 part V B).

The hypothesis regarding the mode of formation of the quartz inclusions is supported

Fig. 74 (No 18). Same as fig. 73, but with roughly trigonal inclusion of mica. In bottom right-hand corner remnant of older mica in cassiterite.

by other observations. In fig. 76, for instance, the outlines of three prisms are recognized, lying in different directions but replaced by one single quartz crystal. If a slide were cut perpendicularly to the plane of this thin section the result would show the same trigonal inclusions as described above. Moreover, fig. 75 shows the cracks and rents along which quartz reached the tourmaline inclusions in the cassiterite. Consequently, the final conclusion

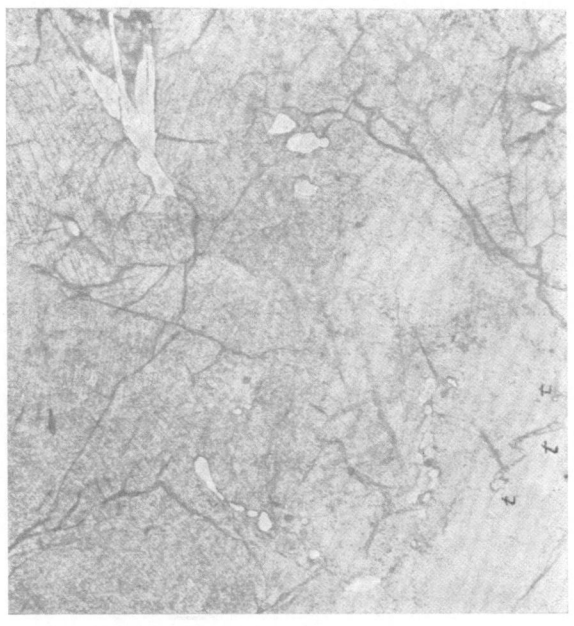

Photo 70 (No 89 Kashozo, magn. 16x, ordinary light). Twinned cassiterite with roughly trigonal quartz inclusions and near right-hand edge a few tourmaline prisms embraced by cassiterite. In top left-hand corner remnant of older mica, cassiterite having been deposited along the homogeneous mica boundaries.

Fig. 76 (No 25). Remnant of cassiterite in vein quartz (white dots) with rows of inclusions. Quartz deposited along cracks in cassiterite and along homogeneous boundary between cassiterite grains A and B. Presumably three prisms of tourmaline (1, 2 and 3) formerly embraced by cassiterite have been replaced by one quartz individual. Quartz originating from crush (Qu II).

is that tourmaline prisms enclosed by cassiterite were subsequently replaced by quartz migrating along cracks in the cassiterite crystals. But, it might be asked, why had the tourmaline not been replaced by mica previous to the deposition of cassiterite? The answer is that probably coarse mica has not been so strongly developed near the present roof as it has near the floor. This explains why tourmaline remnants are still found in the roof quartz and why the upper part of the roof is totally devoid of coarse mica, only some fine-scaly mica being intergrown with the quartz.

Fig. 75 (No 25). Cassiterite (dotted) with inclusions and veinlets of vein quartz (I).

D: *Topaz.*

This mineral occurs in large crystals, but a study of slides does not give a clear idea of its relations with other minerals, sometimes only topaz being seen, showing its characteristic rows of inclusions, its cleavage, low birefringence and typical refractive index, whilst in other cases only mica is seen, with a grain of topaz here and there, all of which are extinguished simultaneously under crossed nicols. The possibility of one large topaz crystal having been replaced by mica must be admitted.

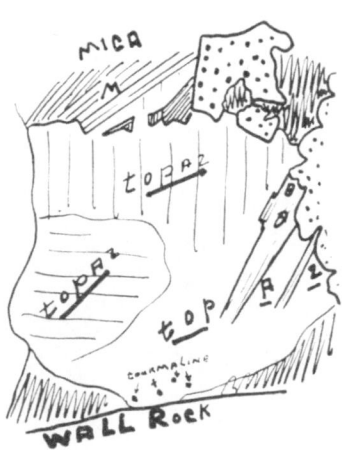

Fig. 77. (Sketched after photo 67). Intergrowth between topaz, cassiterite and mica.

These grains of topaz, however, are not separate grains embraced by mica, but are small veinlets of topaz projecting from the coarse crystal, which fact is revealed by an investigation of the hand specimens. These veinlets follow the homogeneous mica boundaries and their cross sections are mainly bounded by the basal plane of mica. Some of these veinlets are to be seen in fig. 77, which has been sketched after photo 67, and in this figure replacement of mica by topaz along the heterogeneous cassiterite-mica boundaries is particularly evident. Moreover, topaz encloses tourmaline near the floor. Consequently topaz might be younger than mica, tourmaline and cassiterite. None of these minerals occurs along cleavage planes of topaz, nor are they bounded by important crystallographic planes of the latter.

In photo 71 a wedge of mica is depicted. Topaz penetrates a little between two differently orientated folia of mica. Elsewhere (Part V C) it will be seen that topaz is corroded and replaced by quartz, and the fact that at Kaina topaz is totally lacking in quartz seems to bear out the assumption that its time of deposition falls between that of quartz and cassiterite. The latter mineral very seldom occurs enclosed in topaz.

E: *Sequence of deposition.*

In the case of Kaina, therefore, the sequence will be:

1. Sericite recrystallizing to fine-scaly mica.
2. Tourmaline deposited at the cost of the sericite as well as at the cost of fine-scaly mica.

Photo 71. Aggregate of coarse mica partly replaced by topaz (top edge). Photo from unpolished hand specimen. Four differently orientated mica booklets are visible.

3. Mica deposition continues and coarser folia are formed simultaneously with the deposition of tourmaline.
4. Deposition of tourmaline ceases and it is again resorbed by an extremely coarse mica deposited only along the main channel. By this process the vein is formed.

5. Cassiterite is deposited principally in the coarse mica of the vein, but also in or against the wall rock, thereby embracing tourmaline.
6. Presumably there is another short phase during which fine-scaly mica is deposited at the cost of cassiterite and (embraced) tourmaline, but this phase is not of great importance, neither in the vein nor in the wall rock.
7. It is probable that at this stage topaz is deposited, mainly at the cost of coarse mica but possibly also replacing some cassiterite, as the latter only rarely occurs in a few small grains in the topaz.
8. Quartz is then deposited chiefly along the contact at the roof, replacing first of all mica but also cassiterite and enclosed tourmaline. In the country rock, especially in the wall rock at the roof, mica and tourmaline are partly replaced.

This sequence may be tabulated as follows:

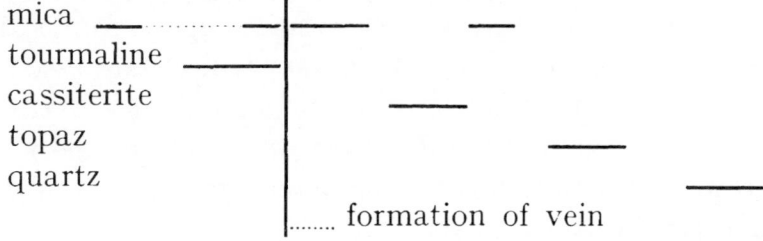

mica

tourmaline

cassiterite

topaz

quartz

.... formation of vein

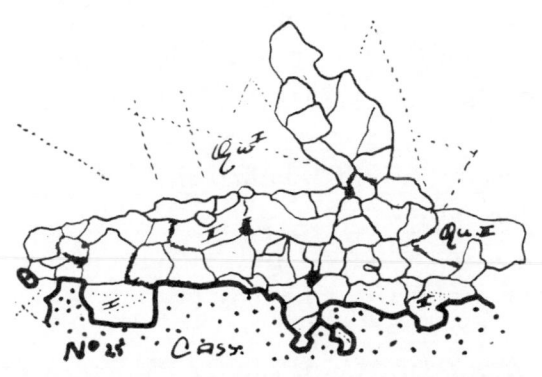

Fig. 78 (No 25) Mode of occurrence of fine-grained clear quartz along cassiterite grains and between two coarse crystals of undulatory vein quartz (with rows of inclusions). Along the grain boundaries of the fine grained quartz are many dark particles.

One more phenomenon remains to be described, and that is the occurence of a fine-grained quartz in the middle of the vein quartz and especially enveloping the coarse cassiterite grains (cf. fig. 78); this sort of quartz has no inclusions, but many dark particles lie along the grain boundaries. This quartz never occurs inside the cassiterite grains in the form of veinlets or inclusions. In Part V B it will be demonstrated that this type of quartz is due to crush, a sort of recrystallization having taken place, in the course of which the inclusions were probably shifted towards the periphery.

The occurrence of this type of quartz is mainly localized along the cassiterite — quartz contacts, which naturally caused inhomogeneities in the rock.

B. STANNIFEROUS DEPOSITS IN PREDOMINANTLY ARENACEOUS ROCKS.

CHAPTER 4

GENERAL INTRODUCTION.

In Part V A a description has been given of a deposit formed in a predominantly aluminous rock. Here the formation will be discussed of a type of deposits which has been formed in more or less pure quartzites and sandstones. The deposits comprised in this part are the stanniferous veins on Buramma Ridge and Ruechimarra Hill lying in the vicinity of the garnet quartzite, and the Kashozo vein (cf. map. 4).

The stratigraphical and tectonical relations of Buramma Ridge and Ruechimarra Hill have already been discussed in Parts II and III, whilst the distribution of additive metamorphism has been described in Part IV B (chapter 6).

The occurrence of stanniferous veins having been explained in Part V, it is not necessary to go into this again (cf. fig. 79).

On Ruechimarra Hill a large number of stanniferous and barren lodes and veinlets have been located in the garnet quartzite, which is entirely devoid of mica.

In many cases their dimensions are extremely small; some of them are in fact so small that they can be taken out in one single hand specimen. They form irregular stock-like pipes or lenses in the quartzite and consist entirely of quartz. These minute stocks (cf. photo 72, No 67) often cut across the bedding or else suddenly make a sharp bend and run along the bedding. No contortion, bending or pushing aside of strata could be observed. Consequently they merely represent irregular shaped alterations in the country rock and the formation of these veins is undoubtedly due to replacement.

Only one lens contained abundant coarse mica, intergrown with cassiterite. It is found on the SW. flank of Ruechimarra hill some hundred metres below the place where the Kigezi boundary joins the road and at an equal distance to the NW. The width of this lode is about 4 ft. and its length about 15 ft.

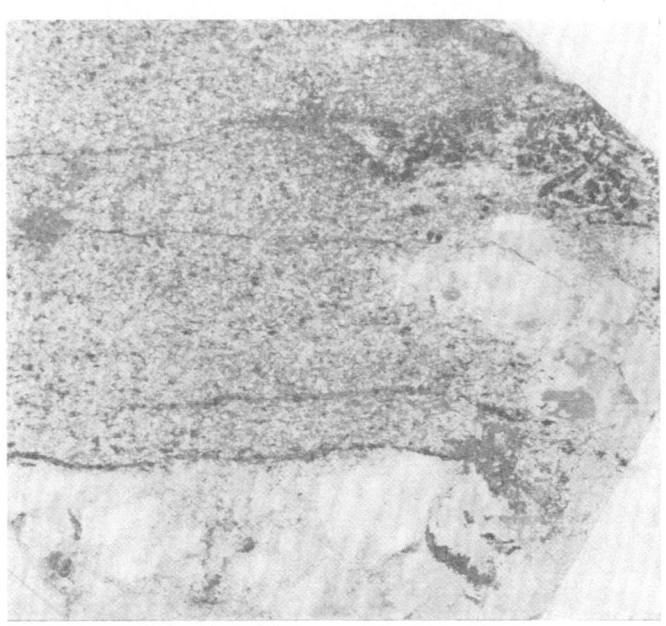

Photo 72 (No 67, reflected light, polished surface, magn. 2x).
Hand specimen of tourmalinized garnet quartzite on Ruechimarra hill with small veinlet of quartz, which is stanniferous,
though no cassiterite is visible in the photo.

This lode cuts across the bedding at right angles. On the
SW. side of the road, in Kigezi
district, several rich but small
lenses were found in or lying
against the garnet quartzite.

More to the SW. on Bu-
ramma Ridge (cf. sketch map
in fig. 79) some big veins have
been found lying in the micaceous sandstone, which un-
derlies and overlies the garnet quartzite. These veins
contain visible cassiterite and
also coarse mica, but mainly
consist of quartz. Their most
important feature is that
they have been considerably
pressed and bent subsequently
to their formation (cf. fig. 80,

lens G), and conse-
quently they show
some abnormal fea-
tures (lenses F and G).

At Kashozo a steep-
ly tilted quartzitic
sandstone occurs in
the vicinity of gran-
ite; it has been strong-
ly impregnated with
tourmaline. In many
parts mica is absent
but in some places it
contains a fair amount
of fine-scaly mica. In
this quartzitic sand-
stone big lenses or
stock-shaped bodies
occur, consisting whol-
ly of quartz. A few
enormous lumps of cas-
siterite were found in-
tergrown with quartz,

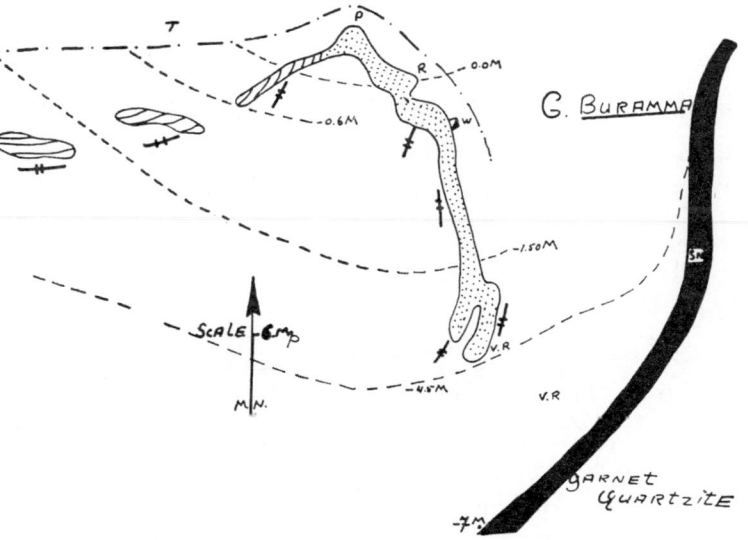

Fig. 80 Sketch map of vein G on Buramma Ridge.

Black   — boundary quartzite
Sn   — cassiterite in veinlet in boundary quartzite
- - - - -   — contours
- · - · -   — upper boundary of detrital cassiterite on surface and
                 in subsoil.
P = poor   VR = very rich (± 3%)
R = rich   w = winze.
The dots indicate the stanniferous, the hatched lines the barren part of
the vein. The dotted and hatched lines give the direction in which the
quartz crystals are elongated (cf. photo 78).
＃ strike (dip is almost vertical). Arrow = magn. north.

but after these had been removed no more cassiterite worth mentioning could be found, in spite of numerous and extensive workings on the outcrop.

The strata abut against these lodes, which in many places have a considerable width, and disappear. Strongly tourmalinized strata are seen to project into the quartz of the veins, but the tourmaline grains soon become more scattered and less numerous. Finally the stratum dies out and this phenomenon proves the metasomatic mode of formation of these lenses, the more so because the strata are not deflected or pushed aside by the quartz veins. All the veins mentioned above are surrounded by metamorphic tourmaliniferous aureoles. These consist of fine-grained, greyish or greenish rocks which still reveal their sedimentary nature (cf. photo 72, and photo 53a, part IV B). The plane of contact between the wall rock and the veins is seldom sharp and clearly defined, with the exception of one single veinlet found at Kashozo. This veinlet of quartz is separated from the metamorphic country rock by a margin consisting solely of long, slender tourmaline prisms, lying on both sides perpendicular to the tabular quartz body of the veinlet.

These tourmaline prisms show signs of corrosion on the side of the quartz; the vein quartz

Photo 73, No 58. Kashozo (ordinary light, magn. 17x) showing coarse, euhedral tourmaline prisms at bottom, with interstitial vein quartz, and xenomorphic tourmaline at top, lying interstitially between the quartzite grains.

penetrates between the prisms or else forms small veinlets in and through the prisms along the basal parting plane. Thus the continuation of the prisms in the quartz becomes irregular and broken. The interstitial vein quartz, and the veinlets along the basal parting are visible in photo 73, which shows the tourmaline margin and the contact with the wall rock. In the middle of the photo the euhedral tourmaline prisms embrace irregular quartz grains and die out. On the extreme right of the photo only a xenomorphic tourmaline predominates, occurring principally along the grain boundaries of the quartzite grains.

In chapter 6, Part IV B, the vein space has been defined as being the space in which by intensive metasomatism such coarse crystals were formed that all palimpsest structures reminding one of the original host rock have disappeared. The veinlet shown in photo 73 is one of the very few instances in which the disappearance of all palimpsest structure is due to the development of coarse tourmaline crystals.

Generally, coarse crystals of tourmaline will not have existed in the pegmatite veins and consequently no traces of their previous existence can be found.

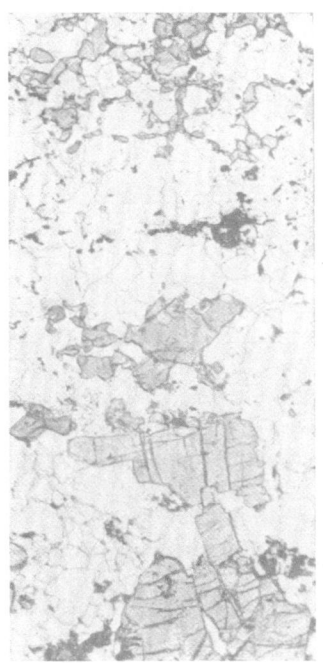

Photo 74. No 85, slide from garnet quartzite on Ruechimarra hill (magn. 16x, ordinary light) taken in the neighbourhood of a small veinlet of quartz and showing the difference in habitus between the xenomorphic tourmaline of the metamorphic quartzite (near top) and the coarse tourmaline crystals of the transitional zone (near bottom edge).

Where it happened to be present, tourmaline has almost entirely been replaced by quartz or mica (cf. photo 51 Part IV B, and 57 Part V of the pegmatitic lens west of hill No 42, Katuba valley). Apart from this, such a sharp contact as shown by the veinlet of photo 73 is rarely encountered and generally the vein is separated from the ordinary metamorphic aureole rock by a narrow transitional zone.

These zones are distinguished by slightly larger dimensions of the grains of tourmaline and quartz. Photo 74 clearly shows the difference in size and shape between the common xenomorphic tourmaline grains in the wall-rock and the irregular crystals in the transitional zone, which mostly are bounded by prismatic faces and the basal parting plane (cf. also fig. 54, part IV B).

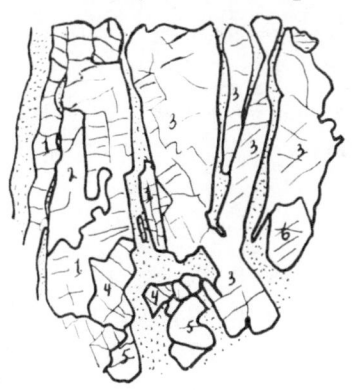

Fig. 81. No 67A Ruechimarra. Cracked tourmaline (white with cleavage; 6 individual crystals) with veinlets of quartz (dotted) along rents occurring in the transitional zone.

Often quartz fills rents in broken tourmaline aggregates (fig. 63). Finally, the presence of more or less parallel, and sometimes continuous rows of crystal and gas inclusions in the quartz is a proof that the quartzite grains have recrystallized under the influence of magmatic solutions (cf. fig. 57, part IV B). These transitional zones differ from the vein proper in that all palimpsest structure has not disappeared. The small tourmaline grains, which in the wall-rock accentuate the sedimentary character of the metamorphic rock, have been entirely replaced in the vein space. In the transitional zone, however, the replacement has only been partial and the incompleteness of this process is typical for these zones.

Photo 72 (No 67, Ruechimarra) shows a twofold magnification of a hand specimen under reflected light. In this specimen a quartz veinlet is seen, first running parallel to the palimpsest stratification and then making an abrupt upward turn before suddenly ceasing. Just above the place where it dies out a fair example of a transitional zone may be observed.

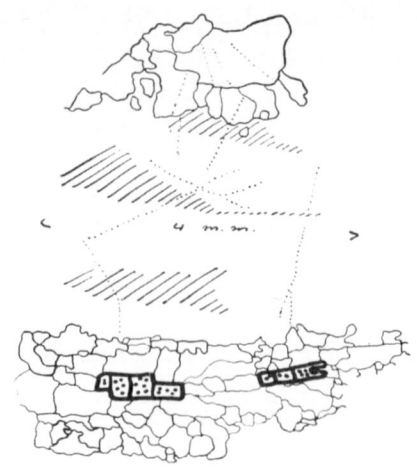

Fig. 82 is sketched after a slide of a veinlet in the same specimen as has been depicted in photo 72. The veinlet shown in this figure is formed by one undulatory quartz individual at least one inch long, 4 mm of which is visible in the sketch. Underneath the veinlet is the quartzite quartz in which two hypidiomorphic tourmaline prisms occur. These tourmaline prisms are almost euhedral in shape but nevertheless it is evident (particularly from the left-hand prism) that they have been deposited by replacement of quartzite grains. The upper side of the veinlet gives a typical example of the characteristics of the transitional zone; in some places rows of inclusions run through from the vein quartz into the grains of that zone; other grains contain rows of inclusions which do not cross

Fig. 82. No 67 Ruechimarra, showing veinlet of quartz consisting of one undulatory quartz crystal in tourmalinized garnet quartzite. Dots: rows of inclusions. Hatched: extinguished parts of undulatory crystal, the white part being bright under crossed nicols.

White: quartz with grain boundaries,
Dotted: tourmaline with heavy outlines.

the boundaries; again other grains are totally devoid of inclusions and a few small grains of this type are enclosed by the vein quartz (in this section at least). In the slide of this sketch large tourmaline crystals as shown in fig 81 (cf. 54, part IV B) are found immediately above this veinlet, and the normal anhedral tourmaline grains, as represented in fig. 53, part IV B, are some distance away from the veinlet. It is, therefore, quite probable that the transitional zone has been built up from quartz grains of various origin, from quartzite grains and vein quartz and from recrystallized quartzite grains.

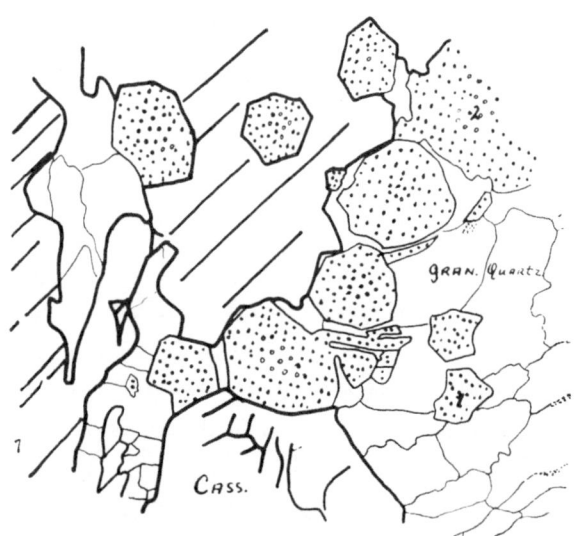

Fig. 83, No 67 Ruechimarra. Transitional zone of veinlet shown in photo 72: Cassiterite (white with cleavage) intergrown with tourmaline (dotted; bluish cores and greenish outer zone) and quartz (white with grain boundaries).

Cassiterite is also found in the transitional zones; fig 83 (also from specimen No 67, shown in photo 72) depicts its mode of occurrence. Probably cassiterite has replaced quartzite quartz, thereby embracing the euhedral tourmaline crystals shown in this sketch.

The tourmaline crystals lie quite irregularly in the cassiterite, and neither their outline, nor their distribution shows the least causal relation to the crystal structure of the cassiterite. It must be noticed, however, that the outline of the tourmaline embraced by quartz is far less perfectly trigonal than the outline of that included in cassiterite.

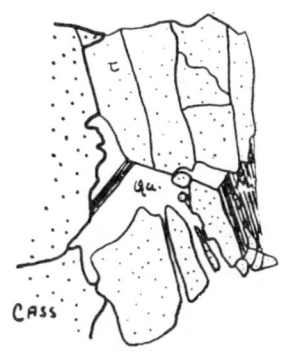

Fig. 84. Same slide as fig. 83. Cassiterite (heavy dots), tourmaline (fine dots), mica (hatched parallel to cleavage) and quartz (white).

Moreover, quartz is seen to occur in small veinlets along the contacts between tourmaline and cassiterite (cf. also fig. 84) and it may be taken for granted that silicification has taken place subsequent to the deposition of cassiterite.

Also at Kashozo cassiterite is present in transitional zones. In the valleys NE and SW of the sandstone outcrop the fragments of cassiterite found in fairly rich detritus had for the greater part orginated from transitional zones, for the cassiterite is intimately intergrown with fine tourmaline, fine-scaly mica and fine-grained quartz. As the transitional zones of the Ruechimarra veinlets contain hardly any mica (the garnet quartzite is totally devoid of it!), the sequence of deposition of the minerals present in the zones will be discussed with the aid of specimens from Kashozo.

CHAPTER 5

TRANSITIONAL ZONE OF KASHOZO VEINS.

Some specimens of this zone have already been discussed in a previous part (Part IV B, i.a. figs. 33 and 34).

Two photos from a slide of this zone (No 38) are given (photos 75 and 76), in which cassiterite, tourmaline, mica and quartz are to be seen. The top left-hand corner of both photos reminds one very much of the aspect offered by slides of metamorphic, silicified and tourmaliniferous mica-schists. Indeed, all the slides from transitional zones of Kashozo veins furnish proof that in these zones a final phase of silicification occurred.

The mica, for instance, is particularly well bounded by its basal plane, and quartz either occurs in wedge-shaped veinlets along 001 into the mica folia or is found on the homogeneous mica boundaries (fig. 85, No 38).

Quartz is also found in tourmaline in veinlets along the basal parting of the latter or on the heterogeneous tourmaline-cassiterite contacts (cf. photo 75, bottom edge), which fact bears out the view that quartz is younger than either of the two other minerals. Microscopical investigation reveals the existence of numerous parallel rows of inclusions in the quartz, often crossing the grain boundaries (see fig. 86, No 38, drawn from near the upper cassiterite grain in photos 75 and 76), and though tourmaline and cassiterite have undoubtedly been deposited for a part in quartzite quartz, the existence of this older quartz can no longer be proved, since through subsequent silicification the quartzite grains have been altered and cassiterite, tourmaline and mica have been partly resorbed.

In chapter 4 of part IV the conclusion has been drawn that cassiterite was younger than mica and tourmaline and that mica already existed at the time of the depo-

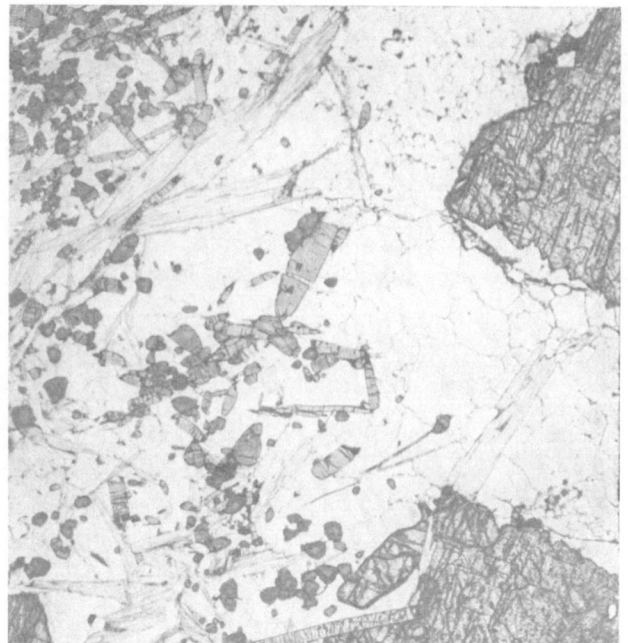

Photo 75. No 38 Kashozo (ordinary light, magn. 16 x). Showing cassiterite (in two right-hand corners), tourmaline (also enclosed in bottom right-hand cassiterite grain), mica and quartz. Notice the nebulae of dark inclusions in the quartz occurring at the ends of several tourmaline prisms.

sition of tourmaline. This conclusion was based on the examination of only a few slides, but in the following pages further evidence will be given to support this assumption. Fig. 87 (No 38), for instance, strongly suggests the idea that cassiterite, which forms veinlets in mica parallel to the basal plane and along the homogeneous contact, is younger than mica. Not the slightest importance is to be ascribed to the presence of cracks in the cassiterite forming the extension of the mica folia. The insignificance of such cracks will once more be demonstrated in fig. 92 in the following chapter.

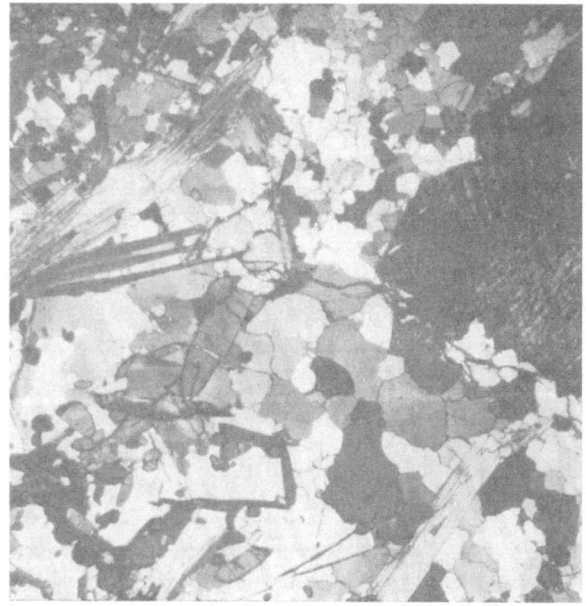

Photo 76. Right-hand part of photo 75 under crossed nicols. Fine specimen of twinning lamellae visible in cassiterite.

Also in fig. 88 the cassiterite is seen to occur along the homogeneous mica contacts. The cassiterite has decidedly an anhedral shape, being bounded by the basal plane of the mica. Quartz has replaced both mica and cassiterite, as will be evident when comparing the cassiterite wedge on the extreme left with the partially resorbed wedge just a little to the right. The deposition of cassiterite along the homogeneous mica contacts is easily seen in fig. 89. Moreover this figure shows the presence of cassiterite along the heterogeneous mica-tourmaline contact, whilst it clearly reveals the fact that tourmaline was not or only very partially replaced by cassiterite. The same may be observed in fig. 90 (No 46). From this last-mentioned figure it appears that the deposition of tourmaline has taken place along the homogeneous mica contacts or parallel to the cleavage plane of the latter. In fact very many similar sketches can be drawn from the Kashozo slides. Therefore the above-mentioned sequence, mica — tourmaline — cassiterite, is undoubtedly right. Cassiterite has preferably replaced mica, but it has merely embraced tourmaline, and very many inclusions of tourmaline are found in the Kashozo cassiterite.

Quartz, too, has replaced mica by preference, whilst many tourmaline crystals are included in quartz, which latter mineral has only partly replaced tourmaline.

The former extension of these tourmaline crystals is marked by clouds of very small inclusions (cf. fig. 85). These inclusions partly consist of gas and partly of crystals The latter are formed by isotropic, isometric grains, which often show good octahedrons and display a violet colour. Possibly they consist of fluorite. The inclusions are found everywhere in vein quartz or

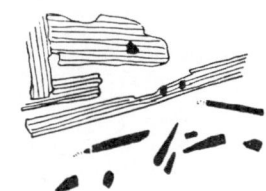

Fig. 85. No 46 Kashozo, showing quartz (white) embracing mica (hatched parallel to cleavage) and tourmaline (black), with nebulae of inclusions (dotted) forming the extension of the tourmaline prisms.

Fig. 86, No 38 Kashozo. Quartz (white with grain boundaries and numerous rows of inclusions) and cassiterite (dotted).

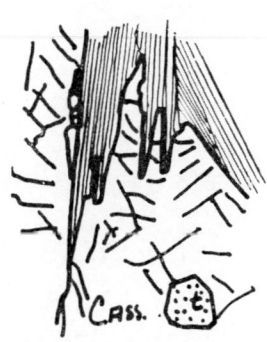

Fig. 87, No 38 Kashozo. Cassiterite (white with cleavage and cracks) embracing tourmaline (dotted) and forming veinlets in mica (hatched parallel to cleavage) along the basal plane and homogeneous grain boundary.

Fig. 88. No 38A Kashozo, showing xenomorphic cassiterite (white with cracks and cleavage) bounded by the basal plane of mica (hatched), both minerals being partly replaced by quartz (white).

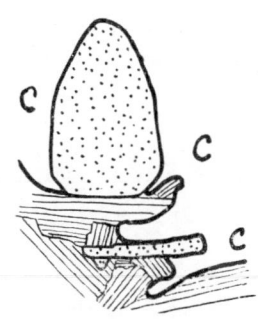

Fig. 89, No 46 Kashozo. Cassiterite (white; "c" embracing tourmaline (dotted) and forming veinlets in mica (hatched).

Fig. 90, No 46 Kashozo — see explanation of fig. 87.

in altered quartzite grains, either in rows or in small nebulae. The nebulae, as already stated, seem to mark the former extension of tourmaline crystals, and their formation or their accumulation is undoubtedly in genetic relation to the resorption of this mineral (cf. fig. 91). These nebulae are always dark, both in ordinary light and under crossed nicols. These very same inclusions are abundantly found in the autometamorphic, pegmatitic varieties of the granite.

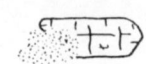

Fig. 91, No 90 Ruechimarra. Tourmaline (with cleavage) embraced by quartz (white), with very small inclusions in the latter.

Whether mica was an original constituent of the host rocks in which these transitional zones of Kashozo veins were formed, or whether it has partly or wholly been added, is uncertain. Here and there the mode of occurrence (cf. fig. 88) seems to point to addition. Nowhere, however, can evidence be found, as in Kaina veins, that mica has been metasomatically deposited subsequent to the deposition of tourmaline and at the cost of the latter.

The sequence of deposition of the hypogene minerals might be as tabulated below:

mica   ————?————

tourmaline   ————

cassiterite   ————

quartz   ————

whilst quartz and mica were probably original components of the host rock.

CHAPTER 6

THE VEIN-FILLING — BURAMMA AND RUECHIMARRA DEPOSITS

The vein-filling of the Kashozo lenses contains only quartz and very little cassiterite. Therefore the sequence of deposition of the minerals in the vein space can best be studied with the aid of slides from the above-mentioned veins, as they contain *mica* as well as *cassiterite* and *quartz*.

A. Cassiterite — mica relations.

The cassiterite occurs in more or less coarse aggregates and big crystals. In nearly every case it is intimately intergrown with mica. From the following it may be seen that cassiterite has replaced mica and therefore the rule deduced from observation that cassiterite is invariably accompanied by mica, but mica not always by cassiterite, is quite comprehensible.

Indeed microscopical investigation has proved that the relations between mica and cassiterite do not differ from those observed in the Kashozo and Kaina lenses. The mica that is intergrown with cassiterite is rather coarse and occurs in the same sort of intergrown booklets as described in connection with Kaina.

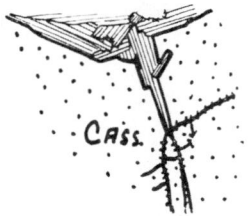

Fig. 92, No 82A Ruechimarra. Cassiterite (dotted, with cracks filled with hematite) and mica (hatched parallel to cleavage).

Fig. 92 (No 82 A Ruechimarra) shows how the cassiterite has been deposited along the homogeneous mica contacts, whilst the heterogeneous contact is remarkably parallel to the basal plane of the mica. The cracks in the cassiterite, which are coloured with hematite, are certainly not due to the preparation of the slide, and their insignificance for the determination of the sequence of deposition is once more demonstrated: it is the cassiterite that shows a preference for the homogeneous boundaries and important crystallographic directions of the mica, whilst the deposition of the latter does not appear to be influenced by any particular direction typical for cassiterite. Undoubtedly cassiterite is younger. The same may be deduced from fig. 93 (No 82 B): along every homogeneous mica contact cassiterite is present. Slides from veins F and G on Buramma Ridge reveal the same feature (fig. 94, No 115n). Most certainly, cassiterite is younger than mica and has been deposited by replacement of the latter min

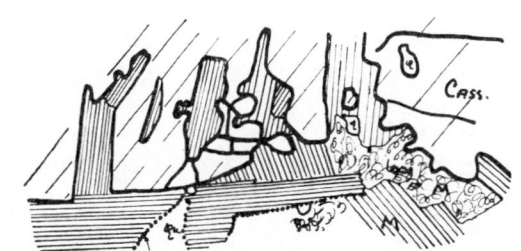

Fig. 93, No 82A Ruechimarra. Cassiterite (white, with cleavage and heavy outline) intergrown with mica (closely hatched parallel to cleavage, or shown by curled marking where cut about parallel to basal plane). Quartz = white.

eral. What, however, will happen when from two centres cassiterite grains begin to grow at the cost of mica? If these centres are lying close together the cassiterite grains will soon come into contact with each other, whilst if the centres are some distance apart maybe a little mica will be left unreplaced between the grains. In fact this is what is seen in figs 95 and 96. For a moment these sketches give the impressions of veinlets of younger mica lying along the homogeneous cassiterite boundaries, but according to the ideas of the author, the fact that the outline of the cassiterite is wholly determined by the crystal structure and homogeneous grain boundaries of mica, already

Fig. 94, No 115n, vein G Buramma. Cassiterite (with heavy outline and hatched parallel to cleavage) and mica (closely hatched). 1 = quartz; 2 = mica.

disproves this suggestion.

Still more convincing is perhaps the absence of mica along the cleavage of

Stheeman, Geology.   9* the cassiterite, the mica veinlets occurring solely along homogeneous grain boundaries.

The author takes it to be an established fact that cassiterite has replaced mica.

B. Tourmaline — mica relations.

In a previous part (part IV, chapter 6) it has already been stated that in the vein-filling small remnants of strongly tour-

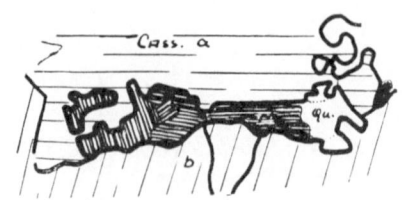

Fig. 95, No 40 Ruechimarra. Remnants of mica (closely hatched) left unreplaced along the homogeneous contact of cassiterite grains a and b (hatched parallel to cleavage). Quartz (white) is present.

maliniferous country rock are met with, consisting of a mosaic of tourmaline grains with interstitial quartz and mica. The probable sequence of deposition has also been discussed and the conclusion arrived at was that tourmaline had been deposited at the cost of quartz and fine-scaly mica, whilst the remaining fine-scaly mica had been partly resorbed by a subsequent silicification (cf. figs 55 and 56, chapter 6, part IV B).

It was considered very probable that a part of the fine-scaly mica had been added to the original sandstone and deposited by replacement of sandstone grains.

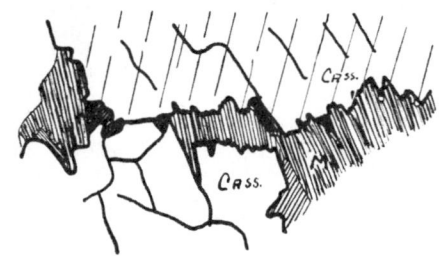

Fig. 96, No 82B Ruechimarra. Same indications as fig. 95:

The coarse booklets of mica mentioned above when speaking of the relations between mica and cassiterite have undoubtedly been deposited by additive metamorphism. It is a very peculiar feature that these folia of mica contain but very little tourmaline. After what has been found from examination of the slides of Kaina veins, it seems to be very probable that the fine-scaly mica has been deposited before the addition of tourmaline took place, whereas the coarse mica might have been deposited after the formation of tourmaline, which latter mineral may have been partly resorbed during this process. This assumption could not be definitely checked as mica is not very abundant in these veins and decisive slides could not be obtained. Figures 97 and 98 (No 90 B Ruechimarra) show the ordinary in-

Fig. 97, No 90B Ruechimarra. Mica (hatched parallel to 001) entirely embracing tourmaline (dotted) and partly replaced by quartz (white).

tergrowth between tourmaline and coarse mica. This tourmaline is sometimes seen lying arbitrarily in the mica, but often it is bounded by the cleavage planes

of the latter. On the other hand, the general rule that tourmaline is restricted to the edges of the mica folia or to the homogeneous mica boundaries does not hold good here, as the tourmaline is often wholly embraced by mica. The author's opinion is that the growing mica folia, which once must have been rather big, have mostly replaced and partly embraced the pre-existing grains of tourmaline. The mica folia were subsequently replaced by quartz, which mineral occurs along the homogeneous and heterogeneous mica contacts or forms veinlets in mica parallel to the basal plane.

Fig. 98. Same indications as fig. 97.

C. The relations of quartz.

Quartz, therefore, has replaced mica, as may be observed everywhere in the veins under consideration. Often several remnants of mica enclosed in quartz have a parallel orientation (fig. 99, No 112 a Buramma G). No veinlets of mica in quartz are

to be observed, but many small veinlets of quartz occur in the mica parallel to its basal cleavage. Tourmaline also appears to have been replaced by the vein quartz, seeing that strongly tourmaliniferous layers of the country rock gradually die out and disappear in the vein quartz. In some veins,

Fig. 99, No 112 n A, Buramma vein G. Remnants of a coarse booklet of mica (hatched) embraced by quartz (white).

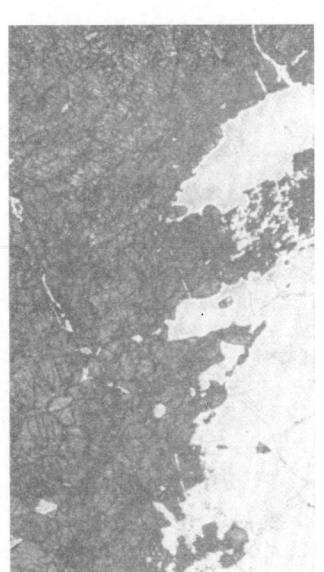

especially those on the Ruechimarra hill, small grains of tourmaline solely bounded by prismatic faces or the parting of the basal plane are often found in the vein quartz, and here again dark nebulae of extremely small inclusions mark the possible former extension of the tourmaline grains. As some of these nebulae

Photo 77, No 115n, Buramma G (ordinary light, magn. 24 x), showing fine intergrowth between quartz and cracked cassiterite.

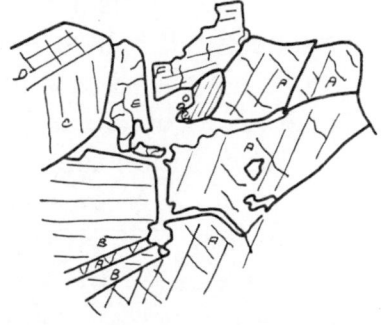

Fig. 100, No 88, Buramma F. Cracked aggregate of cassiterite (white with heavy outline and hatched parallel to cleavage) and quartz (white).

occurring independently in the quartz form hexagons, in harmony with the optical orientation of the quartz, it might be doubted whether these inclusions have *originated* from the replacement of tourmaline by quartz.

Their *accumulation* in nebulae near tourmaline grains is, however, most probably due to the replacement of this mineral. The tourmaline prisms embraced by quartz often show the known crenelated outline described in previous parts.

Cassiterite grains have been intensively replaced by quartz and often one big crystal has been split up into a large number of small grains, all orientated parallel. As many of these veins have suffered from stress, quartz is abundantly present as small veinlets along cracks (cf. photo 58 part V and fig. 100). In figure 100 quartz forms veinlets along homogeneous boundaries of cassiterite while replacing the latter. The diminution of the average dimensions of the cassiterite grains in quartz is due for a great part to this replacement (cf. photo 77), and therefore in order to obtain the same recovery of tin metal, ore containing a large percentage of quartz has to be more finely crushed than ores rich in mica.

CHAPTER 7

CRUSHED QUARTZ

Most of the veins dealt with in this part have suffered more or less from stress and in this chapter the effects of the stress upon some minerals will be described. The presence of crushed and cracked cassiterite crystals has already been mentioned, and especially in the veins F and G on Buramma Ridge (cf. fig. 79) the cassiterite is invariably broken up into numerous small fragments. It is very interesting to note that the cassiterite in these two veins differs greatly from the cassiterite of Kashozo and Ruechimarra deposits in colour and pleochroism: its colour is very dark reddish brown, whilst in polarized, transmitted light it may even become almost dark.

Quartz, however, is the most affected by stress, and photo 78, taken under reflected light from a hand specimen of vein G, Buramma (magn. 2 x), clearly reveals abnormal texture; big, elongated crystals are embedded in a ground-mass of very fine-grained quartz. Under crossed nicols these coarse individuals show a marked undulatory extinction (cf. photo 79, No 115 n, magn. 17 x). They are separated from each other by more or less narrow, triturated zones. In these zones no pseudo-isotropic streaks of finely crushed material are to be found, and undoubtedly to a certain extent recrystallization has taken place. Moreover these small grains of quartz do not show any undulatory extinction. Especially along the inhomogeneous contact between quartz and cassiterite these small grains are of frequent occurrence, and, having originated from crush, in transmitted light they show still another peculiarity: as may be seen in photo 80 (No 88, vein F, Buramma, magn. 16 x), the coarse

Photo 78. Photo of a hand specimen (polished surface, reflected light, magn. 2x) of vein G on Buramma Ridge showing peculiar texture of crushed vein quartz.

Photo 79, No 115n, vein G Buramma (crossed nicols, magn. 17x), showing crushed aggregate of cassiterite and crushed vein quartz. The coarse quartz crystals show an undulatory extinction and are separated by triturated zones.

quartz crystals appear to be more or less greyish with a great abundance of rows of very fine inclusions. The grains of the triturated zones, however, are absolutely clear, except along their grain boundaries.

Fig. 100, No 90A. Embryo of triturated zone between two coarse quartz crystals. Quartz (white with grain boundaries) and rows of inclusions. A small rounded grain of apatite (Ap), which mineral is frequently observed in the vein quartz, is seen at the extreme top right-hand corner.

In fact it seems as if all inclusions have been pushed aside towards the periphery, and such a process must be admitted.

These very same peculiarities are also found in the quartz of the Ruechimarra and Kashozo deposits, though to a lesser extent. Fig. 100 shows an embryonal triturated zone between two coarse quartz crystals (No 90 A, Ruechimarra). The rows of inclusions traverse these zones, making small step faults when crossing a grain boundary.

Some small quartz grains show fewer and others more rows of inclusions than the adjoining grains, whilst yet others are totally devoid of inclusions.

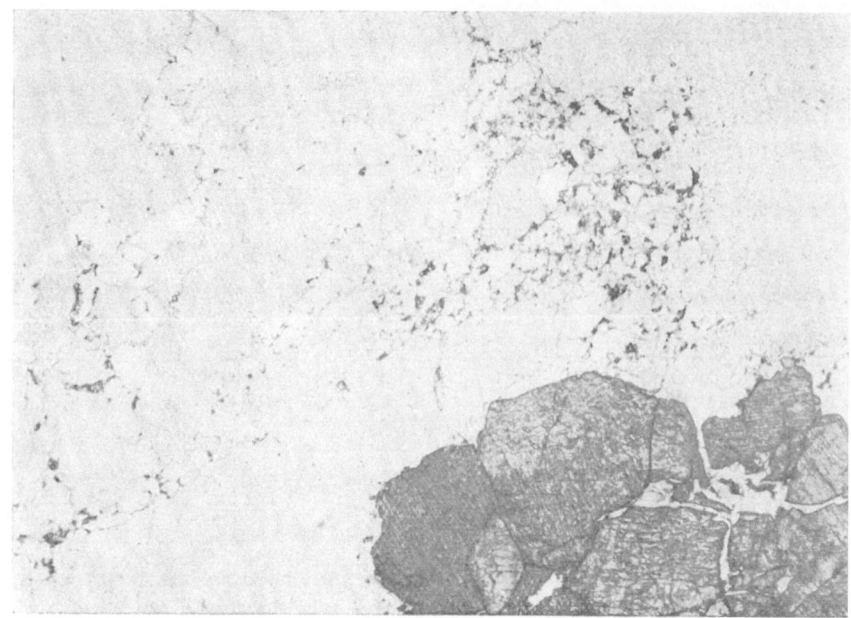

Photo 80, No 88, vein F Buramma Ridge (ordinary light, magn. 16x), showing crushed and partly replaced aggregate of strongly pleochroic cassiterite. Typical arrangement of inclusions in coarse as well as in fine-grained quartz (the latter being clear with many dark particals along its edges).

Indeed signs of crush may often be observed with the naked eye in hand specimens.

C. STANNIFEROUS DEPOSITS IN THE VICINITY OF GRANITE

CHAPTER 8

GENERAL DESCRIPTION

As the heading of this sections implies, these deposits will be discussed irrespective of the composition of the host rock.

Indeed these veins are exactly similar to each other, both in texture and in mineral composition, as well as in shape, no matter whether they are found in sericite phyllites or in the arenaceous phyllites of the Middle Division.

They are almost entirely built up from coarse mica and quartz. On the whole the cassiterite content of these veins was found to be rather low, bulk samples very seldom containing more than ½ %. In the veins near Kabezi, below the boundary quartzite, topaz was present though not abundant (cf. photo 86).

The cassiterite had a deep brown colour, sometimes almost black, as in the pegmatitic vein near the Chamgash (cf. map 4, hill 70). In one and the same vein the proportions between quartz and mica may vary considerably. Two extreme cases are met with: either mica is anhedral in a matrix of quartz, which latter mineral is present in veinlets along all homogeneous mica boundaries and in veinlets in the mica booklets parallel to the cleavage, or separate, angular veinlets and crystals of quartz are found to lie interstitially between coarse rosettes and booklets of mica.

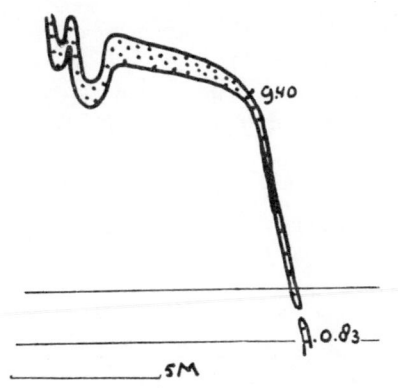

Fig. 101. Section through stanniferous vein near the Chamgash.

These veins generally have a very irregular shape and often enclose large blocks of country rock. They may also occur in large numbers of small, irregular lenses without any visible feeding channel.

In fig. 58, Part V, chapter 1, a series of cross-sections through the vein near the Chamgash are shown. In fig. 101 a diagrammatic section is given of this vein. Its lower part is exactly parallel to the bedding planes of the phyllites, its upper part rather flat, though the influence of the bedding planes is clearly visible. Probably small fractures served as a pathway in addition to bedding planes.

Moreover, from figure 101 it is to be seen that the vein died out in the lower part of the tunnel and appeared again somewhat higher; it did so in both walls of the adit. This might seem rather extraordinary, but the explanation is apparent from photo 81, in the middle of which rosettes and booklets of mica are visible, intergrown with quartz.

Approaching the bottom edge, however, the texture changes and becomes more schistose. This in turn merges into the common country rock, an arenaceous phyllite containing much staurolite and a few needles of tourmaline.

Therefore, these minerals gradually disappear, in the vicinity of the vein, whilst the dimensions of the mica scales increase.

This transitional zone, which has a pronounced schistose character and really consists of metamorphic phyllites with

Photo 81. (Hand-specimen of Chamgash vein, polished surface, reflected light, magn. 2x) showing coarse booklets of mica intergrown with quartz. Near bottom edge transition to contact metamorphic phyllite.

slightly increased dimensions of the constituents, cannot be regarded as vein matter. This rock envelops the vein and constitutes the break shown in fig. 101.

In the vein space of the Chamgash vein some grains of staurolite and tourmaline were occasionally observed; they invariably occurred in inclusions consisting of the transitional schist.

Near Kabezi, too, all tourmaline ceases near the veins, and simultaneously the mica scales attain larger dimensions. Only in the middle of the veins, however, do they occur as really coarse rosettes and booklets.

Undoubtedly staurolite as well as tourmaline have been replaced by one of the vein minerals. Good slides of the contact have not been obtained and the cause of the disappearance of tourmaline is still uncertain. However, in view of what has been noticed at Kaina and on Buramma Ridge it seems fairly certain that the disappearance is due to replacement by mica, the more so as the presence of relatively coarse mica invariably accompanies the disappearance of the abovementioned minerals. *Silicification* occurs in tourmaliniferous country rock as well as in veins, but where *coarse mica* is found tourmaline is absent. Generally some cassiterite is present in this more or less coarse mica.

Often the texture of the mica in the marginal portions of the vein reminds one very much of the texture of the mica in the metamorphic wall rock (cf. photo 55). Consequently it would be obvious to assume that this mica originated solely from recrystallization of phyllitic matter, but this hypothesis cannot be maintained as we are still absolutely in the dark as to whether during the formation of coarse mica addition and removal of material has taken place or not. Consequently it is better to speak of a replacement of fine-scaly mica by coarser mica, with simultaneous disappearance of contact metamorphic minerals.

CHAPTER 9

MICROSCOPICAL INVESTIGATIONS

The relations between quartz, mica and cassiterite are exactly similar to those revealed by slides of the aforementioned veins. As, however, these deposits yielded some very beautiful examples of replacement, this will be gone into further.

Photo 82 is a good picture of the relations between quartz and mica. The former shows a preference for homogeneous and heterogeneous mica contacts, while the latter is often bounded by its cleavage planes.

Cassiterite of a very dark brown colour is found in the mica. The fact that cassiterite has actually replaced mica may be seen from photos 83 a and b.

The shape of two of the cassiterite grains is very uncommon indeed. These grains are absolutely xenomorphic and appear to be pseudomorphic after mica, as is evident from photo 83b, showing in detail a part of 83a under crossed nicols. Looking at 83b alone, one is in doubt which of the two lozenge-shaped grains consists of cassiterite and which of mica. The relations between quartz and mica again suggest a replacement of the latter by the former. The same may be seen from photo 84, where veinlets consisting of fine-grained quartz show a preference for the homogeneous mica boundaries.

This mica is more or less cataclastic, as appears very clearly from a close

Photo 82. (No. 65*a*, Buramma "C", ordinary light, magn. 12x). Cassiterite, mica and quartz.

examination of photo 85: the mica near the right-hand edge is bent and broken;

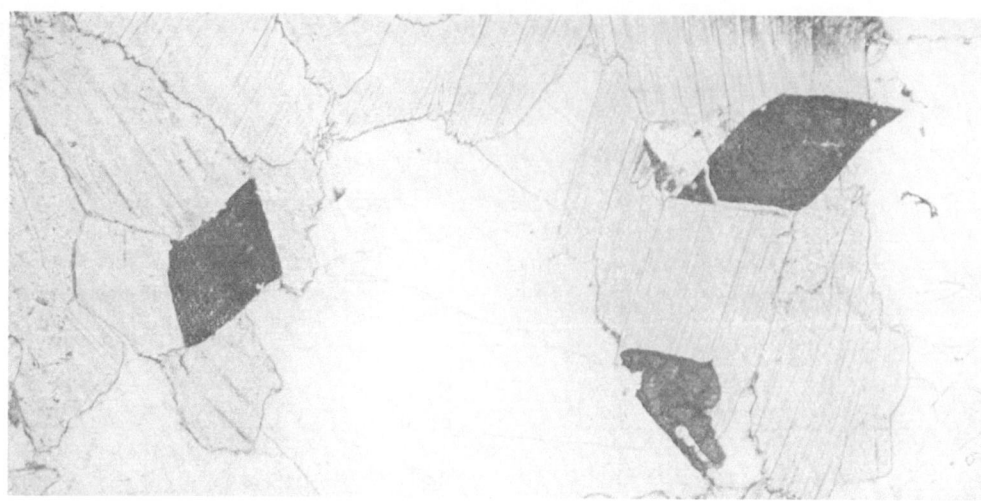

Photo 83*a*. (No. 65*a*, Buramma "C", ordinary light, magn, 18x). Cassiterite, mica and quartz. The left-hand cassiterite grain is evidently pseudomorphic after mica, as is the cassiterite grain in the topright-hand corner.

Photo 83*b*. Part of 83*a* under crossed nicols.

in the left part of the photo, too, near the bottom edge, this cataclasis is apparent. The fine-scaly mica, which lies in the left part of the photo and in which some larger scales are embedded, seems to have originated from crush; it is partly coloured by iron oxides. This

Photo 84. (No. 44, Buramma "C", crossed nicols, magn. 20x). Showing bent and broken mica, partly replaced by quartz.

is the very same sort of mica as has been discussed in connection with photos 62 and 63.

This fine-scaly mica appears to have been recrystallized to a certain extent.

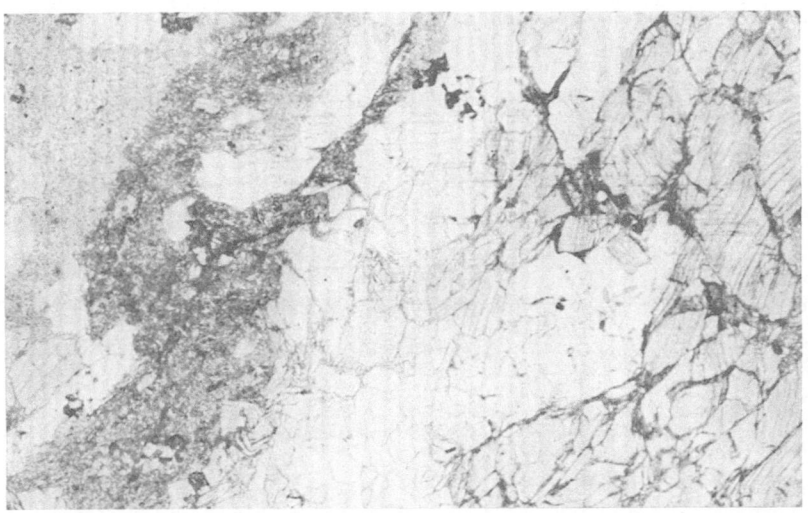

Photo 85. (No. 44*a*, Buramma "C", ordinary light, magn. 16x.) Showing bent and broken mica ((right-hand edge), partly replaced by quartz. Very fine-scaly mica, originating from crush and recrystallized to some extent in left-hand part of photo.

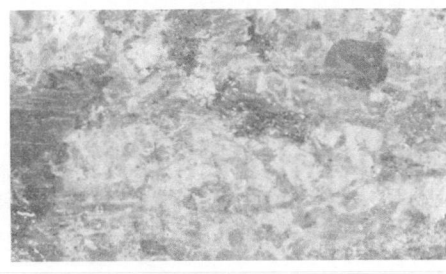

Photo 86. (hand-specimen of vein at Buramma "C", polished surface and reflected light, magn. 2x). Showing coarse, partly replaced booklet of mica (left-hand edge). The milky white specks are irregular topaz grains partly replaced by and included in quartz.

Along the numerous rents and cracks, by which the cassiterite grains in these veins are intersected, a fine-scaly mica is often observed.

Consequently there seems to have been a recrystallization and deposition of fine-scaly mica subsequent to the fracturing of mica and cassiterite. Whether part of this fine-scaly mica has been added or not is unknown.

In photo 86, irregular, whitish specks of topaz are to be seen. This mineral appears to be partly replaced by quartz.

D. CONCLUSIONS.

In every branch of geological science conclusions must of necessity be made with reserve, since the field of observation is limited and as a rule the explanations given for facts observed cannot be checked experimentally. In each of the following conclusions, which the writer has endeavoured to formulate without using too vague terms, such a reserve is, therefore, to be tacitly understood.

1. From the foregoing it will have been seen that subsequent to solidification of the granite there were two main processes, viz. deposition of tourmaline and silicification, the former apparently preceding the latter. Silicification appears to have been accompanied by an abundant deposition of white mica, which, according to microscopical examinations, was formed prior to the deposition of quartz and probably subsequent to the deposition of tourmaline. The latter has to a limited extent been replaced both by mica and by quartz.

2. Tourmaline, white mica and quartz have replaced feldspar, the first two probably having incorporated a few components. F, Bo and silica have been added, whilst alkaline metals and alumina have been removed.

3. The components of the solutions ascending from the granite have already been mentioned in Part V, chapter 1, with the exception of one of the most important, viz. alumina, proof of the presence of which is found in the fact that thick quartzites are sometimes found to have been entirely replaced by tourmaline. Though no analysis of the tourmaline in the contact rocks described is available, yet it may be taken for certain that also iron oxide is an important component and the consequently it was also present in abundance in the solutions right from the beginning.

4. In the sediments a sequence of deposition has been observed similar to that of the autometamorphic minerals: tourmaline is one of the earliest minerals and is followed and partly replaced by mica, with ultimate silicification. Of course, here the deposition of minerals like arsenopyrite and hematite is disregarded.

The comparatively rare occurrence of sulphides and oxides in the Ankolian veins [1]) is a very happy condition, for in many ore deposits one encounters so many minerals which moreover occur in several generations that it is no longer possible to determine the sequence of deposition, but in Ankole this is not the case.

1) Combe reports the discovery of several sulphides, such as galena and chalcopyrite, in a stanniferous vein close to the Ibanda granite. The author believes to have observed chalcopyrite in the form of very minute grains in the metamorphic wall rock of the Kaina quartz reef.

5. Thus the author has been able to establish that the genesis of the ore deposits lies in the deposition of relatively coarse mica resulting in the formation of recognisable bodies in which all traces of the sedimentary nature of the host rock had disappeared.

In many veins this deposition of coarse mica may have been obscured by subsequent silicification, but it may nevertheless be assumed that all the stanniferous deposits previously described have passed through this stage.

In Kaina veins this stage has been definitely proved. In the lenses formed in the immediate vicinity of the granite an increase of the dimensions of the mica folia is to be observed, simultaneous with the disappearance of tourmaline.

Some veins on the Ruechimarra and Buramma ridges contain coarse mica in which tourmaline is either lacking or poicalitically included.

Only in the Kashozo vein is this mica absent, but there *more* relatively fine-scaly mica was found than could possibly have been present in the original quartzitic sandstone; as a matter of fact also the texture of the mica pointed to addition. This mica was older than tourmaline, and in fact in all other veins it has been seen that fine-scaly mica has been deposited prior to the deposition of tourmaline. It is difficult to say whether this mica derived all its components from the country rock or whether it is to some extent of hypogene origin; the former may have been the case in Kaina, whereas the fine-scaly mica found in the tourmaline mosaics in the Ruechimarra and Buramma deposits was probably formed by addition. The deposition of mica has therefore taken a rather long time and tourmaline has subsequently been replaced by coarse mica after first having been deposited in the fine-scaly mica.

6. Cassiterite has been principally deposited at the cost of mica. Undoubtedly it has also been deposited in tourmaliniferous quartzite, thereby replacing the quartz; the percentages found, however, are small and the volume of such ores is negligible. The bulk of cassiterite and the bulk of ore with good percentages is found in micaceous rocks. Subsequent silicification has caused the disappearance, for the greater part, both of cassiterite and of mica. This is the fateful result of silicification, and the thinner the stanniferous mica body the more complete was this effect. Consequently, veins in quartzose rocks in which the initial body consisting of mica was of small dimensions are always poor in tin content.

Aluminous rocks were undoubtedly more favourable for the development of a large and irregular mica body, and as a matter of fact in these argillaceous country rocks it is generally seen that a part of the vein has escaped subsequent silicification. Good values are found therein and consequently a selective mining method is applied. Bulk mining can probably not be applied to low-grade quartzose ores, seeing that the larger the quartz bodies the less cassiterite do they contain.

7. Thus it will be understood that sericite phyllites are a more favourable

host rock for stanniferous veins than the more or less arenaceous phyllites of the Middle Division. Tectonical conditions, however, determine the course of magmatic solutions. A fault, joint or fractured zone attracts large quantities of magmatic solutions, affording relatively easy pathways compared with undisturbed and only slightly permeable sericite phyllites. This may be the explanation of the occurrence of the series of stanniferous deposits along the axis of the Kavungo syncline. The vertical position of the layers, the fractures presumably occurring in the downfold, and the presence of granite in the through bend of the syncline are all factors favourable for the formation of voluminous ore bodies.

8. Finally there is the difficult question whether the erosion of the culminations of the cupolas means a great loss. According to general belief, the smallest arena would offer the best prospects, and this would then be the Rubanda arena. Conditions there, however, are unpromising; tourmaline was not abundant and the only mineral of importance found there, viz. baryte, is characteristic for a low temperature. It is remarkable that there pyrite is of frequent occurrence, whereas in the Ankolian stanniferous veins it is entirely absent.

It may be that the Rubanda granite, for some reason unknown, forms an exception to the general rule that culminations of cupolas offer favourable prospects for the occurrence of metalliferous veins, since father to the NW, where the granite must lie much deeper, both cassiterite and arsenopyrite were encountered in the nucleus of the Bunyoni-Rubanda syncline.

Possibly the discovery of traces of gold in the south of the Kayonza Forest is related to the occurrence of barite and pyrite just referred to.

SW. Uganda forms part of an extensive area in which traces of tin have been found everywhere, and it is possible that at a later date this area will become of more importance. Maybe this paper will then become of more general value, since it is not precluded that the extension of the Karagwe-Ankolian system and its granites coincides with the distribution of tin.

ERRATA.

Page Line

x   Map 4 Scale 1 : 100,000 to read 1 : 150,000

         Sections insert: scale 1 : 25,000

XIV   20   barium to read boron; fluorite to read fluorine

1   1   fair y to read fairly

2   Photo 1 description: Mbara to read Mbarara

3   ,, 3 ,, : Nobugumba to read Nobugamba

7   13   simultancously to read simultaneously

9   30   fig 1 to read fig 17

11   31   section b to read fig 27

13   10   insert "marine" between "its" and "cretaceous"

19   3   these to read those

       10   add "branch" after "the"

21   12   Kirurama to read Kiruruma

22   11   east to read west

25   5   chapter 3 to read chapter 6

29   18   70 to read 75

30   2   Ruigka to read Rukiga

31   10   although to be replaced by whilst

33   Footnote: thickmess to read thickness

35   Footnote, last line: dresent to read present

37   Footnote, 2nd line from bottom: he to read the

43   5th line from bottom: Albertsville to read Albertville

47   Photo 18b: 4a to read 18a

48   11   line to read cline

52   6   these to read the

54   7   syclines to read synclines

55   Fig. 20, in description: fig. 2 to read fig 19

56   36   conpression to read compression

58   Footnote 1: fig 8 to read fig. 25

61   Fig. 25 to read Fig. 23

62   Footnote: last word to read unknown

66   3rd from bottom: fluor to read fluorine; borium to read boron

67   Photo 26b: la to read 26a; Photo 26d: 1c to read 26c

70   Photo 28b: 3a to read 28a

71   1   Fluor to read fluorine

        2   Borium to read boron

73   Photo 32: nothern to read northern; pegmatite to read aplite

79   2   After TiO_2 add "and" in place of the comma

        3   SnO to read SnO_2

83   9   were to read was

      5th from bottom: A. J. Speyer to read A. E. Speyer

86   8   39 to read 38

89   16th from bottom: parellel to read parallel

Page	Line	
99	25	61 to read 59
102	11th	from bottom: after 57 add)
106		Fig. 61 to read Fig. 59
110	16	After 63 add)
111	23	66 to read 64
112		References to fig. 61 to read fig. 60
115	19	ptical to read optical
118	10th	from bottom: delete the second word „time"
121	22	occurence to read occurrence
124	5th	from bottom: "On the extreme right" to read "In the upper half"
125	13th	from bottom: 63 to read 81
131	20	82B to read 82A
135		Fig. 100 to read Fig. 101
	7th	from bottom: 100 to read 101
136		Photo 80, description: particals to read particles
137		Fig. 101 to read Fig. 102
	4th	from bottom: 101 to read 102
138	1	101 to read 102
	27	ditto
139		Photo 82, description: „ordinary light" to read „crossed nicols"
142	22	the to read that
143	13	poicalitically to read poikilitically
144	9	through to read trough
	15	baryte to read barite
	20	father to read farther

Map 4, Explanation: Wishikatwa numbers 8—12 to read 8—24.

    Note: A portion of biotite granite lying east of hill 15 and due north of hill 17 has been left un-coloured.

    Katerero village, lying near the Chobugombe river, SE of hill 15 has not been indicated.

Map 6, Explanation, line 6: aythor to read author;

      „ 9: abreviations to read abbreviations.

    Note: Mt. Nya Kashunzu, situated 1 km. west of Kagoy village in the NW corner of the map, has been omitted.